"Richard C. Smardon, Sharon Moran and April Karen Baptiste have delivered an invaluable handbook on urban stream restoration. Particularly innovative are the book's approach to engagement via arts and culture, and how stream restoration can serve as a way to improve the physical landscape and water quality while at the same time repairing social relationships. It is required reading for urban environmentalists and social justice activists."

—*Jennifer Wolch, William W. Wurster Dean, College of Environmental Design University of California, Berkeley, USA*

"From the nation's capital to the East Bay, from Syracuse to Milwaukee to Chattanooga, Smardon and his colleagues have embraced the rich diversity of experience in revitalizing urban watersheds and waterfronts—a truly collaborative process which will survive the ups and downs of federal environmental policies."

—*Rutherford H. Platt, Emeritus Professor of Geography, University of Massachusetts Amherst, USA*

"This book is a badly needed treatise on how to creatively involve the community in connecting with and caring for the waterways around us, and in the process working towards social justice. This is important reading for any planner, city official, activist or citizen interested in the health and restoration of streams and rivers, and the people and communities they can help to sustain and bring together along the way."

—*Timothy Beatley, School of Architecture, University of Virginia, USA*

"Environmental restoration is a rapidly growing field and the restoration and revitalization of wetlands, streams and other waterways are a dominant focus. In this important book, the authors place their topic squarely within the frames of environmental justice and broadly inclusive public engagement, showing without a doubt that 'success' in restoration is as much a social goal as it is an ecological or technical one. A must read for scholars and practitioners involved in ecological restoration, urban ecology, green infrastructure planning and other fields who are seeking a more harmonious fit between people and natural systems."

—*Paul H. Gobster, US Forest Service, and former Editor-in-Chief of* Landscape and Urban Planning

Revitalizing Urban Waterway Communities

The revitalizing and restoration of rivers, creeks and streams is a major focus of urban conservation activity throughout North America and Europe. This book presents models and examples for organizing multiple stakeholders for purposes of waterway revitalization—if not restoration—within a context of fairness and environmental justice.

After decades of neglect and misuse, the challenge of cleaning up urban rivers and streams is shown to be complex and truly daunting. Urban river cleanup typically involves multiple agendas and stakeholders, as well as complicated technical issues. It is also often the situation that the most affected have the least voice in what happens. The authors present social process models for maximum inclusion of various stakeholders in decision-making for urban waterway regeneration. A range of examples is presented, drawn principally from North America and Europe.

Richard Smardon is a SUNY Distinguished Service Professor Emeritus at the SUNY College of Environmental Science and Forestry, USA.

Sharon Moran is an Associate Professor at the SUNY College of Environmental Science and Forestry, USA.

April Karen Baptiste is an Associate Professor of Environmental Studies and Africana and Latin American Studies at Colgate University, USA.

Earthscan Studies in Water Resource Management

*For more information and to view forthcoming titles in this series, please
visit the Routledge website:* www.routledge.com/books/series/ECWRM/

Revitalizing Urban Waterway Communities
Streams of Environmental Justice

Richard Smardon, Sharon Moran and April Karen Baptiste

With contributions from
Blake Neumann and Jill Weiss

LONDON AND NEW YORK

First published 2018 by Routledge

2 Park Square, Milton Park, Abingdon, Oxfordshire OX14 4RN
52 Vanderbilt Avenue, New York, NY 10017

Routledge is an imprint of the Taylor & Francis Group, an informa business

First issued in paperback 2020

British Library Cataloguing-in-Publication Data
A catalogue record for this book is available from the British Library

Library of Congress Cataloging-in-Publication Data
Names: Smardon, Richard C., editor. |
Moran, Sharon, editor. | Baptiste, April, editor.
Title: Revitalizing urban waterway communities : streams of environmental justice / edited by Richard Smardon, Sharon Moran, and April Baptiste.
Description: Abingdon, Oxon ; New York, NY : Routledge, 2018. |
Series: Earthscan studies in water resource management |
Includes bibliographical references and index.
Identifiers: LCCN 2018001047 | ISBN 9781138698611 (hardback) |
ISBN 9781315474977 (ebook)
Subjects: LCSH: Stream restoration. | Urban hydrology. |
Urban renewal. | Waterways. | Environmental justice.
Classification: LCC QH75 .R477 2018 | DDC 333.91/62153–dc23
LC record available at https://lccn.loc.gov/2018001047

ISBN: 978-1-138-69861-1 (hbk)
ISBN: 978-0-367-60589-6 (pbk)

Typeset in Sabon
by Out of House Publishing

Contents

Figures and tables

Cover

Left: Mosaic depicting Atlantic salmon, Syracuse, New York (see Chapter 9). This Onondaga Environmental Institute project was made possible with funds from the Decentralization Program, a regrant program of the New York State Council on the Arts with the support of Governor Andrew Cuomo and the New York State Legislature and administered by CNY Arts (Photo Amy Samuels)

Right: Sound art project, with people and stream (see Chapter 9). The location is near Onondaga Creek, Syracuse, New York (Photo Fereshteh Toosi)

Centre: Onondaga Creek Walk, Syracuse, New York (see Chapter 7) (Photo Richard Smardon)

Figures

Tables

Author biographies

Richard Smardon is a SUNY Distinguished Service Professor Emeritus at the SUNY College of Environmental Science and Forestry, USA. He has been involved in water resource management for a very long time and has been concerned with how different types of stakeholders become involved (or not) with such applications. Examples include protecting natural riparian floodways, which was the focus of the 1996 publication for FEMA and the Federal Floodplain management council where the dual question was how to avert increased flood damage and protect the more natural features of the floodplain. From 2005 to 2009 Smardon was part of the team to produce the Onondaga Creek Revitalization Plan where the question was how to develop a broad ranging participatory process for urban creek revitalization.

Sharon Moran is an Associate Professor at the SUNY College of Environmental Science and Forestry, USA. Her research interests include the human dimensions of water problems, management of toxic substances, and environmental health policy. She leads the doctoral program in Environmental and Natural Resource Policy (ENRP) in the Graduate Program on Environmental Science. She also teaches classes in environmental policy and governance in the Department of Environmental Studies.

April Karen Baptiste is currently an Associate Professor of Environmental Studies and Africana and Latin American Studies at Colgate University, USA. Her current research has two geographical focuses. In the Caribbean her research explores worldviews to environmental problems with a focus on perceptions of climate change. Her newest research examines the ways in which environmental organizations operate in the Caribbean as an opening to understanding how environmentalism is or isn't part of the decolonization process. Her U.S.-based research examines justice concerns in Central NY as a way to understand how low income residents face environmental justice problems. Her main research interests lie at the nexus of environmental psychology and environmental justice.

Contributors

Blake Neumann is currently pursuing an MS in Water and Wetland Resources at SUNY-ESF, USA, and an MPA at the Maxwell School of Syracuse University, USA. His research interests are in community engagement in resource management and making restoration practice more culturally responsive.

Jill Weiss is a Visiting Assistant Professor in Environmental Studies at SUNY-ESF, USA. Her research interests are at the intersection of conservation and stakeholder communication and collaboration. Prior to her academic work Jill spent 15 years in outreach, research, and strategic planning for environmental organizations in New York City. From 2004 to 2007, she worked with the Bronx River Alliance and its partners to research and create the resource guide *The Bronx River Classroom: The Inside Track for Educators* to strengthen network ties between stakeholders and increase educational and recreational use of the river.

Foreword

> Clean rivers with a wide range of animals and plants have become symbols of a healthy environment, an attractive city, and a stakeholder society having ownership of its environment and responsibility for the needs of future generations.

This summary, in a book published in 2002 with the UK's Environment Agency,[1] placed a strong marker for the revitalization of urban rivers; crucially, a revival based on partnerships with the public and private sectors and the positive involvement of local communities.

A renaissance in urban living was envisaged through a revival that promotes an intimate link between community and nature. With more than 50 percent of the world's population already living in cities—projected to rise to 70 percent by 2050—understanding and responding to the power of urban waterways in the lives of people is essential. But this demands the direct involvement of local communities in the development of both a vision and set of solutions, for the sustainable management of their rivers.

Since the late 1990s, there has been a steady growth in recognition that local people are well-able to work with experts and professionals to produce such outcomes. As colleagues and I discovered in the European SMURF (*Sustainable Management of Urban Rivers and Floodplains*) project (discussed in Chapters 2 and 5 of this book), people living in the heart of urban environments will readily engage with both the emotional and physical criteria that they believe should define their local river. They draw upon history and direct, day-to-day, experience to describe what a sustainable river should be like. Importantly, this is often the opposite of what they currently experience. So—they want a place that they will want to go to, a relaxing, natural, diverse and safe environment that brings a "bit of the rural" into the city.

One of our older participants in the SMURF project, a renowned author on his local river (the Rea in the City of Birmingham, U.K.), drew upon his experience and memory, in the opening of his *Ode to a Secret River*:

When I was young and just a boy,
To play in rivers was my joy,
And as I made my way to school,
I'd dam a stream and make a pool,
Then later on, returning home,
Release the stream and let it roam.

In our community engagement project to restore a stretch of the River Tame in Birmingham, such emotional and physical experiences of being able to once again access the river, rather than being isolated from it by concrete walls and fast flowing water, became essential criteria for its revival. For some this translated into spots where they could get down into the water, for others it was accessible footpaths along the river bank through contrasting areas of light and shade, with abundant wildlife and "babbling water." Importantly, they were able and willing to work with the water engineers, ecologists and local authorities, inevitably constrained in terms of financial resources, to translate what might seem at first sight rather romantic ideals of the river into achievable outcomes. There was willingness to help maintain the river into the future, bringing the next generation through local schools and youth groups, into an understanding of how to sustain a healthy river. Towards the end of the SMURF project, I remember a wonderful Saturday morning when local people of all ages and from diverse communities (including the local Asian community) came out to help plant a wildflower meadow, to take direct ownership of the revival of *their* river.

Community engagement that achieves these outcomes requires time and resources and dedication to the cause. It is far more than a simple consultation exercise, as the multiple examples in this book attest to. But reviving and sustaining urban rivers is worth this investment for the health of cities and all their residents, and for the future. Importantly, as this book stresses, reviving urban waterways is about restoring relationships and community building. In so doing, the authors argue persuasively that there is real potential to tackle issues of environmental justice

Professor Judith Petts, CBE
University of Plymouth, UK

Note

1 Petts, G.E., Heathcote, J. & Martin, D. (2002) *Urban Rivers: Our Inheritance and Future*. IWA Publishing and Environment Agency, London.

Preface

This book was inspired by waterway revitalization processes we've observed in Syracuse, New York, which have encompassed social, economic and environmental dimensions. Our Rust Belt city in Upstate New York is not broadly known for much these days, but its productive past is impressive. Older residents can recite the names of well-known consumer products (e.g., Carrier air conditioners, Syracuse China, Oneida cutlery) once exported nationally and beyond. But many of the factories have now closed, and Syracuse's population has dropped by more than a third. The city has attracted attention for being home to the country's first underwater Superfund site (Onondaga Lake), which is now being remediated. And it has also made the national news for other dubious reasons: Syracuse has one of the highest levels of residential segregation in the U.S., and it is among the worst places to live for African-American and Hispanic people. The rounds of investment, and the decades of disinvestment, have been manifest in the built environments, the ecosystems and the social fabric of the community.

Syracuse is hardly unique. Cities and towns across North America and Europe have experienced similar industrial and cultural upheavals and have lost connection with their natural water resources. Many communities have turned their backs on the streams, creeks, rivers and bayous that once anchored them aesthetically, practically and economically, and, as a result, those waterways have become degraded. Clean air and water are only a small part of what residents have lost through these processes. They have lost their fundamental rights to environmental justice, to access the kind of psychological, social, recreational and aesthetic benefits all people deserve from relationships with the natural world.

But in Syracuse as elsewhere there is reason for hope. Our waterways are taking in much less industrial waste than in the past. Leaders at all levels of government are more committed to environmental justice. As some point out, restoring and revitalizing urban waterways can benefit people and communities in myriad ways, creating more sustainable urban places, reconnecting people with the environment and correcting or at least ameliorating procedural and distributive environmental injustices.

This book explores the concept of such revitalization and looks at examples from around the world. Several terms we use here require clarification. *Revitalize* means making something more alive once again; it is similar to *restore*, but that term implies a more complete putting-back to an earlier state. We settled on revitalize because it is broad enough to cover a range of initiatives and also sufficiently imprecise that it allows us to be inclusive in what we discuss in this book. Also, while restoration projects often focus exclusively on the biophysical integrity of the hydrology and the aquatic ecology, projects revitalizing waterways more frequently target both the waterway and the community or development just beyond it.

The term *urban* seems intuitive, but for the purposes of this book we note that its definitions can take two different forms, and we embrace both of them. The main one suggests a static quality—how large or how dense—and implies a threshold for designation of urban. The other considers process instead of endpoint—urbanizing rather than urban, which matters for waterways because urbanization harms aquatic ecology, even at low levels of impervious surfaces. *Waterway* here includes linear water bodies usually known as streams, creeks or rivers. While it's hard to specify a precise distinction between stream and rivers, differential flow rates may provide an approximate indicator. Still, we note that even the USGS sidesteps any quantitative definition of a river, instead acknowledging that it is relative.

Our goal here is to explore academic scholarship as well as pragmatic applications that center on waterway restoration and environmental justice. We hope this book will prove useful to both academics studying such subjects and to residents, advocates and activists engaged in the long-term processes of restoration and revitalization.

Richard Smardon
Sharon Moran
April Karen Baptiste

Acknowledgments

Much of the genesis of this book lies in work done in connection with planning the Onondaga Creek Revitalization Plan (OCRP), starting more than a decade ago. Working from Syracuse, NY, the authors would like to acknowledge with respect the Onondaga Nation, the indigenous people whose ancestral lands we now stand on. In doing the research for a book of this nature, there are many who contributed their time and effort. We wish to thank Edward Michalenko, Ph.D., Onondaga Environmental Institute Director, who had the foresight to write the grant proposal for the Onondaga Creek Conceptual Revitalization Plan, and Meredith Perreault in her critical role as project manager. Other key contributors included Lee Geches, Samuel Gordon, Sid Hill, Tarki Heath, Amy Samuels, Samuel Sage, Jeannie Shenandoah and Kelly Somerlot. This was truly a collaborative effort of five different organizations plus a 15-member citizen-working group.

Learning about the best practices taking place nationally and beyond was an essential and compelling part of planning the course of action for the OCRP. For the Bronx River case study, we would like to thank for their insight members of the Bronx River Alliance; Sustainable South Bronx; Steve Oliveri of the Ministry of Peace and Justice; Linda Cox, Executive Director of the Bronx River Alliance and NYC Parks and Recreation; Joan Byron, Senior Fellow at the Pratt Institute Center for Community and Environmental Development; and Jill Weiss, former Bronx Educational Program Director. We would also like to thank Gerry Willis of the National Park Service, who provided valuable context.

For the Wildcat-San Pablo Creek case study, we would like to thank Alan LaPointe and Anne Riley for their insight.

For the Guadalupe River, San Jose, California, case study, we would like to thank Kathy Muller for her insight as the first director of the Friends of Guadalupe and Leslee Hamilton (subsequent director). Also we thank Al Gurevich, Unit Manager for the Guadalupe River Watershed, for his frank appraisal of ecosystem-based management approaches.

For the South Platte River Greenway, we thank Jeff Shoemaker, Executive Director of the Greenway Foundation, for his time and insight regarding the South Platte Greenway development. Jeff provided some of the impetus for the Onondaga Creek revitalization back in 2000. We also thank Bar Chadwick, former chair of the South Platte River commission, for her insight regarding long-range planning.

For the Milwaukee River Greenway, we would like to thank Cheryl Nenn of Milwaukee Riverkeeper, Andrew Kurth and Kimberly Gleffe of the River Revitalization Foundation, and Urban Ecology Center for their insights. Also special thanks to Ann Brummitt, Milwaukee River group coordinator, who came on board in 2006 to work on the Milwaukee River Greenway Plan. In addition, photographer Eddee Daniel and Alderman Nik Kovac provided valuable context.

For the Chattanooga, Tennessee, River Greenway case study, we would like to thank for their insights Jim Bowen, who worked on the initial planning for the greenway, as well as David Unruh with the River City Company; Jim Brown with the Tennessee River Gorge Land Trust; Philip Grymes, Executive Director for Outdoor Chattanooga; and Rick Wood with the Trust for Public Land in Chattanooga.

We would like to acknowledge Professor Anne Spirn formerly of the University of Pennsylvania for her dedication and active scholarship regarding the Mill Creek in Philadelphia. We should also note the work of Rutherford Platt, Emeritus Professor at the University of Massachusetts, for doing early spadework on urban waterway revitalization and surfacing many of the key social issues involved.

We would like to acknowledge the foundational work of Professor Judith Petts at Plymouth University, U.K. and Dr. Ulrika Aberg of Eawag—the Swiss Federal Institute of Aquatic Science and Technology—for their work for REstoring rivers FOR effective catchment Management (REFORM) with regard to participatory process development, implementation and evaluation. Also Mark Everard's work on the valuation of ecosystem services as part of urban waterway revitalization has been foundational.

We would also like to thank the numerous student research assistants who have been part of the project. At SUNY-ESF, graduate students Blake Neumann and Lauren Vermilion helped with research, and at Colgate University, Sabrina Callender-Clewett and Sydney Bird (both class of 2019) were integral to some of the secondary data collection and analysis for parts of this book.

In terms of funding, we thank Colgate University's Research Council for supporting some of the publication costs of this book.

We would like to thank Earthscan acquisition editor Tim Hardwick for bringing us on board with our publisher, as well as Amy Louise Johnston for guiding us through the manuscript submission process.

We would like to acknowledge the graphic assistance of Ryan Mackerer and Mark Warfel Jr., former landscape architecture students, who did all the maps and line drawings. Conversion of color images to black-and-white halftones was done gratis by Industrial Color Labs in Syracuse, NY.

We especially owe a debt of gratitude to Harriet Brown, Professor of Magazine Journalism at the S.I. Newhouse School of Public Communications at Syracuse University, for her careful and timely editing of our text.

1 Introduction

Urban waterway history and planning context

Richard Smardon

Introduction

We know that negative impacts from urbanization accumulate within watersheds as small tributary streams contribute higher peak flows and lower base flows to waterways downstream. As Platt points out, "Dozens (and possibly hundreds) of small urban watersheds in the United States and around the world ... are the focus of multifarious "restoration" strategies under complex institutional arrangements" (Platt, 2006, p. 29). In many instances, as with our own Onondaga Creek in Syracuse, New York, the main channel has been straightened and "hardened," moving higher flows faster through the settled areas without flooding. These same channels in many cases are lined with sanitary sewer overflows (SSOs) and combined sewer overflows (CSOs), which may dump raw sewage plus street drainage during storm events, thus severely degrading the urban waterway.

Many obstacles can interfere with the revitalization or restoration of urban waterways and what follows are examples of such obstacles. The urban creeks, streams and sloughs in need of revitalization are often within poor neighborhoods with highly diverse populations and across multiple jurisdictions. Some examples are Wildcat Creek in North Richmond/San Pablo, California (Riley, 1989a, 1989b); the Anacostia River in Washington, D.C. (Powell, 2010); and Onondaga Creek (Moran et al., 2013) in Central New York State (Figure 1.1). Local communities may not agree as to what should be done to revitalize these waterways; different agencies may hold conflicting priorities, e.g., flood control vs. water quality improvement vs. habitat restoration.

An incredible number of research and demonstration projects (Bernhardt et al., 2005) have attempted to restore segments and functions of small creeks, streams and bayous, but we maintain that a major challenge for the urban waterway restoration/revitalization is gaining consensus about what to do and how to do it. We learned this during the three years of working on the Onondaga Creek Revitalization Plan (Moran et al., 2013) in Syracuse, New York, and others have found this to be a major challenge as well (Moran, 2003, 2007, 2010; Platt, 2006; Riley, 1998). There are also

Figure 1.1 Onondaga Creek walk linking Franklin Square to Downtown Syracuse
Source: Photo by R. Smardon

equity issues in terms of who has historically been forced to live in high-risk floodplain or polluted water areas.

The term *restoration* implies speaking mainly from a biophysical restorative functional capacity, e.g., hydrology, water quality, aquatic and riparian habitat. Using the term *revitalization* implies social and economic improvement or revitalized creek neighborhoods with economically sustainable land use patterns as well as some level of biophysical restoration of the water body. *Naturalization* implies some degree of biophysical water body restoration.

The focus of this book is to explore social processes that are equitable to surrounding communities and can be combined with good environmental science to advance urban waterway restoration/revitalization. The few other books that address the subject include Ann Riley's *Restoring Streams in Cities: A Guide for Planners, Policy Makers and Citizens,* published in 1998, and the American Planning Association's *Ecological Riverfront Design: Restoring Rivers, Connecting Communities* (Otto et al., 2004). One of this book's co-authors produced and co-wrote *Protecting Floodplain Resources: A Guidebook for Communities* for the Federal Interagency Floodplain Management Task Force (Smardon et al., 1996), and the book includes both participatory and agency driven processes for protecting riparian areas within floodplains.

The following sections of this chapter will review the history of urban waterway restoration, U.S. government programs, state and regional stream restoration activity and urban waterfront revitalization, and will make the case for equitable social models for urban waterway restoration.

History of stream restoration innovations and contributions

According to Riley (1998), one of the earliest documented stream restorations involved trees planted along water channels dug in the Euphrates and Nile River valleys in 3000 B.C. Water plants were used for nature-based recreation in both Persian and Roman cultures in about 1000 B.C. The Mayan civilization in Mesoamerica implemented water canals and water retention structures for food production and transportation circa A.D. 1000 (Smardon, 2006, 2009). Water features were used in the Mediterranean cultures of North Africa, Spain and Italy circa A.D. 1200.

According to Riley (1998), stream engineering techniques were first developed in Europe in 1662. Planners moved to using water in landscape design in the 1600s to 1800s. Italian landscape designers utilized decorative water features in garden design, which spread to France and England (1600–1700). English landscape designers used streams, lakes and ponds in pastoral landscape design (1700–1800).

Frederick Law Olmsted, America's first landscape architect, incorporated urban waterways into garden and greenway designs in places like the Fens in Boston and the Genesee River Gorge in Rochester, N.Y., between 1860 and 1880 (Fabos et al., 1968). Later, village and rural improvement societies and women's clubs developed beautification programs that started in New England and spread west to California (Riley, 1998).

Environmental awareness of the use of water in landscape design surfaced in George Perkins Marsh's book *Man in Nature*, which addressed human impact on the environment, including streams. In 1870, the American Fisheries Society was formed and started fisheries restoration work in streams. Teddy Roosevelt assembled Congress, governors, scientists and outdoor sportsmen in 1908 for the White House Conference on Natural Resources, which begat both the Natural Resources Commission headed by Gifford Pinchot and the North American Conservation Conference with Canada, the U.S. and Mexico in 1909. In 1914 the Hetch-Hetchey was a proposed dam site in Yosemite Valley when activists like John Muir spurred the beginning of the U.S. environmental movement. In 1922 the Izaak Walton League, a group long concerned with maintaining high-quality fishing streams and conservation in general, was founded in the U.S.

The evolution of single use to multiple uses of streams and rivers began in the 1930s, when Hugo Schiechtl pioneered soil bioengineering techniques and the Institute of Fisheries Research was established at the University of Michigan for stream habitat improvement research (Riley, 1998). W.C. Lowdermilk (Assistant Secretary of the Soil Conservation Service) toured Europe around the same time and discovered stream restoration techniques such as Schiechtl's. In 1933 Aldo Leopold published his book *Game Management* explaining the concept of carrying capacity, which could be related to riparian habitat areas.

Gilbert White's 1942 dissertation *Human Adjustments to Floods* attributed flood losses to acts of humankind and led to work on flood management as a field (Platte, 1986; White, 1960, 1969). In 1964, Phil Lewis at the University of Wisconsin proposed multiple-use river corridors, which led to the greenway concept. Ian McHarg published *Design with Nature* in 1968, which proposed physical design determinants as a series of mapped overlays to guide future development and resource use. Roy Mann also published his book *Rivers in the City* in 1968, which celebrated attributes of rivers in European and North American cities. The NGO American Rivers was founded in 1968, and continues today as a major advocacy organization for river protection and restoration.

The 1980s brought the revival of stream restoration in North America. The President's Commission on America Outdoors was held in 1987, and Charles Little's 1990 book *Greenways for America* signaled the beginning of the greenway movement in the U.S. The Coalition to Restore Urban Waters was established in 1993 and pushed for a National Watershed Restoration Act, which was not passed by Congress until 1996. In 1998, the Federal Interagency Restoration Working Group (which includes all major federal agencies concerned with water management) produced *Stream Corridor Restoration: Principles, Processes and Practices*, a compendium of all types of stream restoration information. That same year Ann Riley also published her book *Restoring Streams in Cities: A Guide for Planners, Policy Makers and Citizens*.

The 1990s brought urban ecology into the fold of stream restoration as well with Anne Spirn's book *The Granite Garden*. This more holistic systems approach stressed the notion of ecological services provided by streams, including water purification, flood attenuation, microclimate moderation, nutrient flows and aquatic and riparian habitat maintenance (Everard & Moggridge, 2012). The 1990s also saw increasing use of Rosgen's geomorphic stream classification, which can be used to rebuild streams in a more natural pattern without causing harm from downstream flows. In 2005 the National River Restoration Science Synthesis (NRRSS) databases were created, which synthesize 37,099 river and stream restoration projects (Bernardt et al., 2005).

Rosgen, with little formal training in restoration science, promoted his natural channel design (NCD) classification and restoration process (Rosgen, 1994) throughout the U.S. from the 1980s to the present. His NCD process has been adopted by many U.S. federal agencies such as the U.S. Environmental Protection Agency, the Natural Resource Conservation

Service, the U.S. Fish and Wildlife Service and the U.S. Forest Service. Some research scientists have criticized Rosgen's approach as too formulaic but have not offered an alternative other than more detailed studies. The so-called "Rosgen wars" have been documented by Rebecca Lave (2012) and Lave et al. (2010).

The 1990s to the early 2000s saw the advent of the urban stream day-lighting movement, promoting the physical daylighting of buried urban streams to allow sunlight access and biophysical improvement of water quality and habitat attributes (American Rivers, n.d.).

This is a brief history of stream restoration innovations in terms of concepts, publications and other contributions. The following section will look at government programs that supported or hindered stream restoration/revitalization.

Historical chronology of government programs

The 1800s to early 1900s in the U.S. saw the early jurisdictional development of U.S. government agencies, which were beginning to be more and more involved with water resource management. In 1802 Congress established the U.S. Army Corps of Engineers (USACOE) to do surveying and military construction work. In 1824 Congress gave the USACOE a civil works mission involving waterway navigation management, and in 1879 the Mississippi River Commission was established to address basin-wide navigation and water management issues.

By 1898 the U.S. Forest Service had actually started watershed conservation practices and the Bureau of Reclamation was established with the 1902 Federal Reclamation Act. As previously mentioned, Teddy Roosevelt convened the Conference on Natural Resources in 1908, which begat the Natural Conservation Commission in 1909—the first national agency to address natural resource conservation in coordinated fashion. In 1927 Congress expanded the USACOE jurisdiction with the Rivers and Harbors Act and there was limited acceptance of responsibility for flood control with this measure.

The 1930s through the 1940s saw the creation of Franklin Roosevelt's New Deal programs, which in turn empowered agencies to address major natural resource issues. In 1933 the National Industrial Recovery Act marked the beginning of New Deal programs, and in 1934 Franklin Roosevelt appointed the Natural Resource Board. In 1935 the Emergency Relief Appropriations Act was passed. Together these measures allowed public works projects, massive job creation and many water resource development projects.

Many agencies became much more involved with natural resource conservation activities during this time. The U.S. Forest Service developed bio-engineering handbooks in 1933, and the Soil Conservation Service (SCS) was established within the U.S. Department of Agriculture to develop soil and water conservation projects primarily in agricultural and rural areas. In 1935 the Tennessee Valley Authority (TVA) was created to handle water supply, flood control, navigation and power development—all within one agency. In the same year the Bureau of Reclamation was made part of the federal public works program mentioned earlier. In 1936 the Flood Control Act gave the USACOE responsibility for flood control for all major river basins. Congress also passed the Wildlife Restoration Act, which gives states assistance to acquire and restore wildlife habitats. From 1933 to 1944 New Deal programs created natural resources planning organizations, including comprehensive works programs for river basin plans and studies plus coordination of USACOE, TVA and the Bureau of Reclamation. These programs created jobs for 8 million people with $13 billion to restore and/or utilize land and water resources.

The 1940s to the 1970s have been called the Golden Age of Channelization (Riley, 1998) as these newly endowed federal agencies created 34,140 miles of channelized streams with 1,630 separate projects. This was also a time of competing agency jurisdictions and controversial decision-making about which projects got funded and how. In 1941 the USACOE started to develop standards for efficiency analysis for water resources projects, which later became known as cost benefit analysis. In 1942, Gilbert White's dissertation, as noted earlier, found that flood losses were due to human actions, a finding that led to the development of the field of flood management (White, 1957, 1960, 1961, 1966, 1969). This plus the publication of *The Flood Controversy* by Luna Leopold and T. Maddock fed the controversy over effectiveness of flood control projects for both the SCS and USACOE.

Agencies' missions were expanded to address flood control and other water resource issues. In 1956 the Small Watersheds Act put SCS in the flood control business with limited-size structures for small watersheds. In the 1960s SCS expanded services to incorporate conservation measures, including land use planning, erosion, runoff sedimentation and water quality maintenance measures.

By 1960 the USACOE was administering a nationwide flood management program, and the National Flood Insurance Program was established. This program gave residents and businesses insurance against flood damage, provided these communities met minimum land use planning measures and/ or used flood proofing for structures sited within the 100-year floodplain. By 1973 the Flood Disaster Prevention Act required projects with land use controls within the 10-year flood way, thereby narrowing the benefits allowed.

From the 1960s to the 1980s publicly funded water resource projects increased by 40 percent, but flood damages also rose and public support for channelization was weakening because of undesirable environmental and social impacts from these projects. With the passage of the National Environmental Policy Act of 1969, water resource development agencies had to become more accountable for environmental impacts and disclosure of these impacts.

The 1980s to the 1990s saw movement from single-purpose to multiple-purpose water resource projects as well as a struggle to address environmental and social issues within water resource projects (Kusler & Larson, 1993). In 1973 the U.S. Water Resource Council (WRC), which coordinated federal water resource agencies, implemented Principles and Standards for Project Planning, which made national economic development and environmental quality co-equal in terms of project decision-making. In 1979, an executive order created the Federal Emergency Management Administration (FEMA) to coordinate federal disaster assistance. In 1980 the House Committee on Science and Technology asked the National Science Foundation for program recommendations to improve natural flood damage reduction. Also in 1980 the WRC's Principles and Standards were rescinded by the Reagan administration, and the USACOE cost sharing for the non-federal share increased from 25 percent to 35 percent or more, with a 50 percent cost share for feasibility studies. This was a major shift in project cost sharing that made some water resource development projects unaffordable for poorer communities (Riley, 1989a) and so constituted an environmental injustice with regard to access to federal government resources. At the same time the SCS became the Natural Resource Conservation Service (NRCS) and became active with project planning and implementation in smaller urbanized watersheds such as in San Jose and Chicago.

In 1985 the Farm Bill-Food Security Act (the so-called Sodbuster Act) created the Conservation Reserve Program offering farmers and ranchers incentives for wetland protection and eliminated support for agricultural drainage of wetland areas (Smardon, 1992, 2009). The Water Resource Development Act (WRDA) of 1986 again shifted cost benefit analysis for project feasibility to greatest net benefits and required environmental quality benefits to at least be equal to project cost.

The 1990s saw a revival in watershed-wide restoration planning with changes both in floodplain management programs and water resource development programs (Kusler & Larson, 1993). In 1990 there were incentives built into the National Flood Insurance Program via the Community Rating System (CRS) for reducing flood insurance premiums from 5 to 15 percent by going beyond minimum land use/floodplain standards. By 1993 the Hazard and Relocation Assistance Act, which was a response to the 1993 Mississippi River Flood, created $45 million for acquisition and relocation of properties within floodplains with 25 to 75 percent cost sharing.

In 1990 the Food Security Act put limitations on the percentage of water resource projects for urban areas for NRCS, which had developed the reputation of being easier to work with on such projects than the USACOE. The 1990 WRDA directed the USACOE to develop an environmental mission plus a wetlands action plan with a no net loss goal. In 1992 the WRDA directed the USACOE to integrate restoration into dredging projects.

From 1993 to 1996 the Coalition to Restore Urban Waters pushed for a Watershed Restoration Act, which would have provided funds for river, stream and wetland restoration. In 1995 the USACOE restoration project guidelines were issued, and the 1996 WRDA removed barriers for initiating nonstructural responses with 35 to 65 percent cost sharing. This same WRDA also included provisions for aquatic ecosystem restoration and protection projects and a separate provision for stream restoration projects, which have been very popular throughout the country. Rosgen's (1994) geomorphic stream classification system has been used to dechannelize or repair damaged streams to make them more natural, though as we stated before there is some disagreement about the application of the natural channel design approach utilized by Rosgen (Lave, 2012; Lave et al., 2010).

In 1998 the Interagency Stream Restoration working group published *Stream Corridor Restoration: Principles, Process and Practice*, a very large compendium of information about stream repair. The same group, led by NRCS, is still working on an even more detailed technical publication to complement the 1998 publication (Fripp et al., 2003). The 1990s also saw many communities developing stormwater management plans as dictated by the U.S. Clean Water Act Phase II stormwater management. This was passed, by the U.S. EPA, to the states and local jurisdictions for implementation. As part of these plans many communities are now considering use of best green management practices (BMPs) to reduce urban runoff rates and volumes, which in turn will help streams reduce peak storm flows.

This is a brief history of federal agency programs that have affected stream restoration and/or revitalization. Those wanting more details should see work by Riley (1998) and Platt (1986) as well as Vileisis (1997) for wetland management program history. The take-home message is that poorer communities were more likely to sustain more flood damage and risk and less likely to receive aid for urban waterway restoration or revitalization.

State programs and regional stream restoration activity

There is much activity in stream restoration/revitalization around the country. There are also a number of synthesis papers that cover this activity by Bernhardt et al. (2005), Bernhardt & Palmer (2007), Carpenter et al. (2003), Gracie & Clar (2003), Schwartz et al. (2003) and Platt (2006). According to Palmer et al. (2003), these are the dominant types of restoration projects drawn from the NRRSS databases by region:

- Northwest: Project related to habitat rehabilitation for better salmon fisheries.
- Upper Midwest: Stream bank stabilization and salmonid habitat improvements.

- Chesapeake Bay: Stream bank grading, matting/live stakes and riparian planting, stormwater management, bank stabilization.
- Southeast: Bank stabilization, channel reconfiguration, riparian planting.
- Central U.S. large rivers: Channel widening, channel/floodplain reconnection, sand bar/island reconstruction, flow management.
- Southwest: Native species conservation, exotic species management, endangered species recovery, flow regime management.
- California: Channel reconstruction and levee alteration, gravel augmentation, fish screens and habitat improvements, riparian planting and fencing, bank stabilization in urban channels.

Carpenter et al. (2003) reviewed regional preferences and accepted practices in urban stream restoration projects. These most often include reintroducing meanders to straightened stretches using a natural channel design approach, channel daylighting of closed conduit streams, bed and bank stabilization, dam and culvert removal and riparian habitat removal. Key background information techniques can be found in the *Stream Corridor Restoration Manual* produced by the Federal Interagency Stream Restoration Working Group (1998), the USACOE Hydraulic Design of Stream Restoration Projects (Copeland et al., 2001) and the Urban Stream Assessment (Brown, 2000).

The following is a regional stream work overview gleaned from Carpenter et al. (2003).

Pacific Northwest: The major driving factor is salmonid migration, but stream restoration projects also address water quality, fish and wildlife habitat, resident endangered species and bank stabilization (Booth, 2005). There is much urban waterway activity in both the Portland and Seattle urban areas.

California is a very active region for stream restoration work because of the multitude of active watershed conditions. Plus, California has the Urban Streams Restoration Program administered by the California Water Resources Department. This program evolved from urban stream projects addressing flooding in the mid-1980s (Riley, 1989a, 1989b). The San Francisco Bay Regional Water Quality Control Board has published a technical reference on stream protection (Riley, 2002). Then of course there is Ann Riley's 1998 book on urban stream restoration, which evolved from California stream restoration projects (Woolley et al., 2002; Ortiz, 1989). The Wildcat Creek in North Richmond/San Pablo and the Guadalupe River in San Jose, California urban stream restoration projects are covered in this volume as reference case studies as well as the Los Angeles River (Wessels, 2010; Wolch, 2007).

Southwest: As noted by Carpenter et al. (2003), the hydrology of most of the southwest is quite different from the rest of the country. Arid conditions and occasional torrential rains create difficult factors affecting restoration design. According to Carpenter et al. (2003), there is not much stream restoration activity in the southwestern U.S. compared with other regions.

Midwest: There seems to be a focus on restoring natural processes that create and maintain habitat rather than manipulating in-stream habitats. There is little standardization guidance, but many state agencies are utilizing stream classification and assessment guidelines. Colorado, Missouri and Kansas have guidelines for screening criteria for stream restoration feasibility (Carpenter et al., 2003). The outstanding urban example for this region is the work going on with the South Platte River greenway in Denver, which will be treated as a reference case study within this volume.

Great Lakes region: Consistent issues here are degradation of urban watersheds due to modified hydrology and morphology, reduced water quality and altered riparian vegetation. Often within the region, habitat enhancement activities differ between warmwater and coldwater species. Examples of stream restoration projects in the Great Lakes region include Mill Creek in Highland Hills, Cuyahoga County, Ohio, and the North Branch of the Chicago River in Illinois. The case studies representing this region will be the Milwaukee River Greenway in Wisconsin and the Chicago River (Gobster & Westphal, 1998, 2004; Otto et al., 2004).

Southeast: North Carolina and Georgia have moved forward with implementing urban stream restoration projects, and South Carolina, Kentucky, Alabama and Tennessee are all developing urban stream restoration projects and protocols (Carpenter et al., 2003). Arkansas is also taking a holistic watershed approach. The Chattanooga Tennessee River stretch will be a reference case study as well as the adjacent Chattanooga Creek (Rogge et al., 2005).

Mid-Atlantic: Most of the activity in this region is in the Chesapeake Bay-Maryland area, where a ten-year history of stream restoration projects has been documented by Brown (2000). Projects in this region fall into four categories: bank protection techniques, grade control structures, flow deflectors and bank stabilization techniques. Many Maryland state agencies have become involved in these projects. In Pennsylvania the Keystone Stream team and the Alliance for Chesapeake Bay published guidelines for natural stream channel design for Pennsylvania Waterways (Keystone, 2003), and in Virginia local soil and water conservation districts have initiated stream restoration activities. The major mid-Atlantic case study is the Anacostia River in the Washington, D.C. area (Arnold et al., 2014; Haynes, 2013; Powell, 2010).

Northeast: In this region we have a long history (300 years) of river and stream channelization, dams and dikes for land drainage, water power and flood control. Former uses of old mills, canals and dams have become uneconomical, thus opening up riverfronts for reuse. There is increased pressure for recreation, fish passage and improved water quality for many northeastern water bodies. Massachusetts initiated its River Restore Program in 1999, and the first dam was removed in coastal Massachusetts in September 2002. Both Vermont and New Hampshire are addressing flooding and erosion issues of high-gradient streams as well as fish habitat and ecological management issues. There has been some stream restoration work in Connecticut with Trout Brook in West Hartford and Piper Brook in Newington.

In New York State, potential river restoration work is either planned or ongoing in several urbanized watersheds, including the Buffalo River in Buffalo, which is actually an Area of Concern (AOC) with a Remedial Action Plan under the Great Lakes Water Quality Act. Both Onondaga Creek in Syracuse, N.Y., and the Bronx River in New York City (Campbell, 2006; Hopkins, 2005) are in the process of revitalization with some restoration work and are covered as reference case studies in this volume. A stream restoration guidance publication called *Stream Processes: A Guide to Living in Harmony with Streams in New York State* (Thigpen, 2006) was produced in Chemung County, New York.

Waterfront revitalization projects

Other activities related to streams and urban water bodies are waterfront revitalization projects, which address more of the socio-economic aspects of waterway restoration. Such projects emphasize revitalization as an economic engine for tourism development (Wrenn et al., 1982), heritage preservation (HCRS, 1980) and/or greenway recreational access development (Little, 1990; Fabos & Ahern, 1996). One of the earliest examples is the San Antonio Riverwalk in Texas, which was designed as a major economic tourism generator (Figure 1.2). Reviews of urban waterway revitalization projects can be found in the publications of the NPS/Association of State Floodplain Managers (NPS/ASWM) (1991), the National Park Service (NPS) (1996), The Urban Land Institute (Wrenn et al., 1982) and USDI Heritage Conservation and Recreation Service (1980). There is also the American Planning Association's *Ecological Riverfront Design: Restoring Rivers, Connecting Communities* (Otto et al., 2004), which documents the revitalization work on the Chicago River. From these publications a few examples of river-related waterfront revitalization projects have been culled and are presented in brief below:

Figure 1.2a and b San Antonio Riverwalk
Source: Photos by R. Smardon

The Charles River in Boston and Cambridge, Massachusetts, is mentioned by NPS/ASFM (1991), Platt (2006) and Spirn (1984) as an outstanding use of an urbanized river corridor. It preserves 8,000 acres of wetlands in the Charles River watershed as an alternative to upstream dams and levees

Figure 1.3a and b Charles River Esplanade
Source: Photos by R. Smardon

while maintaining floodwater control. The project also preserves wildlife habitat and gives urban residents recreational access. From Frederick Law Olmsted's time (Fabos et al., 1968) until today, the Charles River corridor has been a vibrant part of the Cambridge-Boston quality of life (Figure 1.3).

The South Platte River in Denver, Colorado, is mentioned in many reviews for its waterway revitalization (HRCS, 1980; Platt, 2006; Spirn, 1984). Today the Platte River greenway is generating new residential and mixed-use riverfront development while maintaining multiple functions of flood storage, water quality enhancement, riparian habitat and urban recreational amenities in the center of Denver (see Figure 1.4).

The Chattahoochee River in Atlanta, Georgia, maintains a natural river corridor and integrates conservation and development within the growing metropolitan area of Atlanta. Innovative aspects include land vulnerability standards, river buffer zone and floodplain standards plus voluntary protection zones (NPS/ASFM, 1991; Smardon et al., 1996).

The Kickapoo River project in Soldiers Grove, Wisconsin, involved moving an entire business district out of the floodplain and reconstruction of the business center as a solar-powered village. Homes on the floodplain fringe were raised above the 100-year flood elevation. Relocation brought economic revitalization as well as reduced hazard from flooding (NPS/ASFPM, 1991).

Figure 1.4 The South Platte in Denver
Source: Photo by R. Smardon

Boulder Creek in Colorado brought about a transformation in the creek through channel improvements, boulder structures, boat/fish ladder/diversion dams to improve fish habitat and whitewater boating while maintaining irrigation diversions and flood capacity. Agreements were made with irrigators to allow maximum stream flow for fish habitat while protecting water rights (NPS/ASFM, 1991).

Wildcat-San Pablo Creek in North Richmond, California, is one of the earliest stream restoration projects in California. An advocacy planning process empowered local citizens during a multi-objective planning approach. A government-citizen team developed natural stream channel designs that maintain ecological characteristics and capacity to convey a 100-year flood (NPS/ASFM, 1991; Riley, 1989a, 1989b). This is one of the earliest cases where regional environmental and academic experts worked with a low-income community to give them a voice in the process of revitalization (see Figure 1.5).

At Mingo Creek in Tulsa, Oklahoma, in response to a series of damaging floods, the City of Tulsa developed a plan for a system of greenways and trails linking multiple-purpose flood control structures along Mingo Creek (NPS/ASFM, 1991; HRCS, 1980).

Figure 1.5 Wildcat Creek in Richmond CA
Source: Photo by R. Smardon

Thanks to citizen action from the Save the American River Association, the American River Parkway in Sacramento, California, today hosts myriad activities on its 23-mile-long, 5,200-acre greenway. Three major county parks plus a dozen smaller parks offer opportunities for boating, rafting, tubing, hiking, cycling, picnicking, day camping, golfing and nature study. A special overlay zone helps control inconsistent development (HRCS, 1980).

In Savannah, Georgia, a 30-acre esplanade runs three-quarters of a mile along the Savannah River between the water's edge and the rows of old warehouses that have in turn been turned into restaurants and shops for tourism. Local businesspeople played a key role by initiating action to rehabilitate parts of the waterfront to create a catalyst for major economic development at the water's edge (HRCS, 1980). Right behind the river waterfront are areas where historic slave trading occurred, which is not reflected in the local narrative, so this is a historically environmental justice riverscape (see Figure 1.6).

The Trinity River runs through the metropolitan area of Dallas-Fort Worth with 240 acres of floodplain. The river corridor is the largest remaining tract of open space with riparian habitat and historic areas within the region. Major flooding in 1989, 1991 and 1995 spurred action

(a)

(b)

Figure 1.6 a, b and c The Savannah Riverwalk
Source: Photos by R. Smardon

Figure 1.6 (Cont.)

to create the nation's largest cost-shared interjurisdictional, multiple-purpose USACOE study. Results include the Dallas Trinity River Corridor Citizens Committee recommendation for a $17.3 million Phase 1 plan for (NPS, 1996):

- Corridor development certificate process for flood minimization.
- Dalhoma implementation plan for 120-mile multi-jurisdictional, multi-purpose corridor linking Dallas and Oklahoma.
- Regional flood warning implementation plan; and
- Greenway and Trinity Trail Implementation Plan with $13 million in IESTEA Transportation Enhancement funds.

The need for social process models

Given the different waterway objectives for restoration and/or revitalization, there is a need for a collaborative social process with social equity for developing such projects and plans (Moran, 2003, 2007, 2010). For urban waterways such as rivers, creeks, sloughs and bayous that fall within multiple jurisdictions and affect diverse stakeholders, this is especially challenging as pointed out by Platt (2006), Riley (1998) and Smardon et al. (1996). So the question is where are the social process models for urban waterway restoration and/or revitalization? A linked issue is social equity—who has a voice as a stakeholder within such processes? We maintain there is very little guidance in this regard if one looks at the most accessible sources.

Within the first six chapters of her book *Restoring Streams in Cities: A Guide for Planners, Policy Makers and Citizens* Ann Riley lays out the roles of urban river planners, environmental professionals, river scientists and hydraulic engineers. Chapter 8 in her book, "Citizen-Supported Restoration Activities," includes data collection, water quality monitoring and water-shed inventories, checklists for identifying waterway management needs, and watershed and stream management for property owners.

Within the publication *Protecting Floodplain Resources: Guidebook for Communities,* Smardon et al. (1996) include Section 4, "Planning for Resource Protection & Restoration," which lists the following six steps:

- Identify the planning area.
- Conduct an inventory and analysis of land use and environmental concerns.
- Conduct a problem and need assessment.
- Define corridor management boundary.
- Develop an action plan/agenda, and
- Implement and monitor the action plan.

Any of these steps can be done by a government agency, a private nonprofit or a public–private partnership.

The *Stream Corridor Restoration: Principles, Processes and Practices* handbook developed by the Federal Interagency Stream Corridor Restoration Working Group (2000) includes about 16 pages (out of 300-plus pages) called "Getting Organized and Identifying Problems and Opportunities." This section covers setting boundaries, forming an advisory group, establishing technical teams, identifying funding sources, establishing points of contact and a decision structure, facilitating involvement and information sharing among participants, and documenting the process. To be fair there are sections throughout the handbook that could involve public participation, but most of the book is technocratic in orientation.

So the dilemma is that there is substantial "technical information" about urban waterway restoration/revitalization, but not much social guidance for involving multiple stakeholders in the process with varying levels of knowledge about waterway restoration/revitalization. The following outlines some of the major social issues involved with such projects and plans.

From our experience with the Onondaga Creek Revitalization Planning (OEI, 2009) effort in Central New York, the watershed-wide planning process transcends political and economic stratification of a metropolitan region, including both rural and urban stakeholders and even a sovereign nation (Moran et al., 2013). Building public awareness of rehabilitating a long-neglected urban creek can foster a sense of place and community, as others have found (McGinnis, 1999). The Onondaga Nation, in the middle of the Onondaga Creek watershed, certainly had a deep symbolic and spiritual relationship to the Onondaga Creek and Onondaga Lake watershed.

As Platt has stated (2006), the Federal Water Resources Planning Act of 1964 started with engineering water resource development projects, which have evolved to a broader range of goals, means and stakeholders for watershed management. According to Platt (2006) and others, "ecosystem management" is becoming the dominant paradigm for resource managers, and the renegotiation of public versus private responsibilities is leading to decentralized power arrangements with collaborative processes. Such arrangements are facilitating new citizen participation strategies (Cortner & Moote, 1994; Daniels & Walker, 1996), including collaborative partnerships and examples of "collective action institutions" (Lubell et al., 2002) and ecosystem-based management (Daniels & Walker, 1996).

As the process opens up there are greater possibilities for debate over goals and means for stream restoration (NRC, 1992), and such processes may cause conflict between engineering-based approaches and naturalized stream restoration costs. This in turn may cause uncertainty, delay and greater transaction costs (McGinnis et al., 1999; Eden & Tunstall, 2006; McDonald et al., 2004). One example is the Guadalupe River in San Jose, California (see Figure 1.7), where a very expensive bypass system was installed to alleviate peak flow flooding. The local water authority with Friends of Guadalupe River instituted an adaptive planning process, and now California water authorities are pushing for a more naturalized reconfiguration of the Guadalupe River, causing a great deal of anguish for the local water authority. Portions of the greenway and park system are used by homeless populations, causing another management issue that was not envisioned in the design process.

Legal mandates since 1970 have provided tools and funding for creative watershed programs, but as Platt (2006) and Moran (2007) point out, these may actually exacerbate conflict. The conflicts could involve private property owners and public recreation interests, developers and wetland managers or different groups of recreational users, e.g., motorized versus non-motorized. One example is the Milwaukee River Greenway, which is trying to institute a greenway overlay zoning setback that would include a height restriction for multiunit apartment structures and office buildings adjacent to the Milwaukee River (see Figure 1.8). Clearly, this is a potential conflict between land developers and public amenity interests.

Given the social participatory process challenges just itemized, there is little work in Europe or North America that provides practical or theoretical process guidance. We have social processes for revitalizing urban waterways in Europe (Eden & Turnstall, 2006; Eden et al., 2000; McDonald et al., 2004; Petts, 2006, 2007; Tunstall et al., 2000) and North America (Carrol & Hendrix, 1992; Daniels & Walker, 1996; Kondolf, 1998, Lubell et al.,

Figure 1.7 Guadalupe River in San Jose California
Source: Photos by R. Smardon

Figure 1.8 The Milwaukee River upstream
Source: Photo by R. Smardon

2002; McGinnis, 1999; McGinnis et al., 1999; Walker et al., 1999). Much of this work emphasizes place-based and co-production processes between local residents and professionals (Laison & Lach, 2008; Petts, 2006, 2007; Wessels, 2010). Such social processes that address equity issues will be threaded throughout this book.

Plan for the book

Our plan for this book is to explore the potential of learning through deliberative process (Petts, 2007) and collaborative learning models in general (Daniels & Walker, 1996) with social equity. *Revitalizing Urban Waterway Communities: Streams of Environmental Justice* will present basic models for organizing multiple stakeholders for purposes of waterway revitalization or naturalization, if not restoration. This book will focus on environmental justice issues and use the Onondaga Creek Revitalization Planning process as the central organizing tool but will also include other case studies throughout North America and Europe. Key case studies will include the Anacostia River in Washington, D.C.; the Bronx River in NYC; the Milwaukee River Greenway; the Tennessee River in Chattanooga; the Wildcat-San Pablo Creek in Richmond, California; and the South Platte in Denver. The objective is to find common lessons learned from successful and unsuccessful participatory processes for river, creek and stream revitalization within urban areas and for addressing environmental justice issues. The following is an overview of the chapter contents for this book:

Chapter 2, **History of urban river restoration in Europe**, focuses on the U.K. with other European countries. This chapter includes a policy and science context of urban river restoration/revitalization in Europe and will be drawn from Boon et al. (1992, 2000), Boon & Pringle (2009), Boon & Raven (2012) and Petts et al. (2002), plus European urban case studies scattered where appropriate in the succeeding chapters.

Chapter 3 contains a **big-picture framing of the challenge we are interrogating**: stream restoration initiative communities with environmental justice issues. This chapter includes a review of two lines of thinking: *political ecology* and *environmental justice*.

Chapter 4 includes **environmental justice issues of leadership and intergenerational continuity**. This emphasizes the role of demographic groups and organizations arrayed against environmental justice issues. This chapter is mildly theoretical and focuses on unlikely combinations of populations such as the Onondaga Creek Conceptual Revitalization Plan with the rural areas to the south, the Onondaga Nation, and the urban intercity population. Case studies include the Anacostia River near Washington, D.C.; Chattanooga Creek in Tennessee; and Mill Creek in Philadelphia.

Chapter 5 looks at **public engagement and forming partnerships**. It reviews what has worked and what has not. It highlights what has been done with various kinds of groups (scouts, school programs, local teachers, etc.). It also emphasizes how programs that fit into existing structures and incentives—for instance, teachers and continuing education requirements—have a greater possibility of success. We review different types of partnerships such as the Milwaukee River with its four different organizations and the Tennessee River in Chattanooga with its four different functioning organizations versus the South Platte with one main driving organization.

Chapter 6 makes the point that by changing the frame we are changing the relationship and that **restoring streams is really about restoring relationships**. This could specifically be a flavored take on the points on the Native American engagement with the Onondaga Creek Conceptual Revitalization Plan. Broader issues of environmental justice issues are also addressed. Stories and examples like Onondaga Creek, Wildcat Creek and the South Bronx are covered. The key for this chapter is using creek revitalization as restoring relationships and community building and/or revitalization.

Chapter 7 engages **community mapping**. It includes the whole range of activities that have been done with maps as catalysts for engagement. It could include the by-hand-on-a board kind of map, the use of participatory GIS and whatever people are doing with online and social media and community mapping and watershed projects. The Onondaga Creek charrette process of using interactive maps and playing cards is the main example and is supplemented by others.

Chapter 8 covers urban creek revitalization as part of **green infrastructure** for urban areas with a particular emphasis on urban resident physical and psychological health improvement. This is particularly key as a way of compensation for past environmental justice situations.

Chapter 9 engages **creative and artistic based engagement strategies**. In this chapter we discuss the uniqueness of places and solutions, and have an opportunity to roll out engaging stories that illustrate the creativity of communities, including the Golden Ball event to celebrate the Bronx River revitalization work. The theme here is place-based engagement and there are many examples of this within the case studies.

Chapter 10 is a **final summary** or "where should we go from here" chapter. This chapter summarizes lessons learned and projects that need to be done to address participatory process and environmental justice issues.

References

American Rivers. (n.d.). *Daylighting Streams: Breathing Life into Urban Streams and Communities*. Washington, DC: American Rivers.

Arnold, C.A., Green, O.O., DeCaro, D., Chase, A. & Ewa, J-G. (2014). The socio-ecological resilience of an eastern urban-suburban watershed: the Anacostia River basin. *Idaho Law Review*, *51*, 29–90.

Bernhardt, E.S., Palmer, M. A., Allan, J. D., Alexander, G., Barnas, K., Brooks, S., Carr, J., Clayton, S., Dahm, C., Follstad-Shah, J., Galat, D., Gloss, S., Goodwin, P., Hart, D., Hassett, B., Jenkinson, R., Katz, S., Kondolf, G.M., Lake, P.S., Lave, R., Meyer, J. L.,

O'Donnell, T.K., Pagano, L., Powell, E. & Sudduth, E. (2005). Synthesizing U.S. river restoration efforts. *Science, 308,* (5722), 636–637. doi: 10.1126/science.1109769

Bernhardt, E.S. & Palmer, M.A. (2007). Restoring streams in an urbanizing world. *Freshwater Biology 52,* 738–751. doi: 10.1111/j.1365-2427.2006.01718.x

Boon, P.J., Calow, P. & Petts, G E. (1992). *River Conservation and Management.* Chichester, UK: John Wiley & Sons.

Boon, P.J., Davis, B.R. & Petts, G.E. (2000). *Global Perspectives on River Conservation: Science, Policy and Practice.* Chichester, UK: John Wiley & Sons.

Boon, P.J. & Pringle, C.M. (2009). *Assessing the Conservation Value of Fresh Waters: An International Perspective.* New York, NY: Cambridge University Press.

Boon, P.J. & Raven, R.J. (2012). *River Conservation and Management.* Chichester, UK: Wiley-Blackwell.

Booth, D. (2005). Challenges and prospects for restoring urban streams: a perspective from the Pacific Northwest of North America. *Freshwater Science* 24 (3), 724–737 [online] https://doi.org/10.1899/04-025.1

Brown, K.B. (2000). *Urban Stream Restoration Practices: An Initial Assessment.* Elliot City, MD: The Center for Watershed Protection.

Campbell, L.K. (2006). Civil Society Strategies on Urban Waterways: Stewardship, Connection, and Coalition Building. (Unpublished Master's thesis). MIT, Cambridge, MA.

Carpenter, D., Slate, L., Schwartz, L., Sinha, S., Brennan, K., … MacBroom, J. (2003). Regional preferences and accepted practices in urban stream restoration: An overview of case studies. In M. Clar, D. Carpenter, J. Gracie & L. Slate (Eds.), *Proceedings from ASCE 2003: Protection and Restoration of Urban and Rural Streams* (pp. 128–142). Philadelphia, PA: American Society of Civil Engineers.

Carrol, M. & Hendrix, W.G. (1992). Federally protected rivers: The need for effective local involvement. *Journal of American Planning Association, 58(3),* 346–357. doi: https://doi.org/10.1080/01944369208975813

Copeland, R.R., McComas, D.N., Thorne, T,R., Soar, P.J., Jonas, M.M., … Fripp, J.B. (2001). *Hydraulic Design of Stream Restoration Projects.* Vicksburg, MS: Coastal Hydraulics Laboratory, USACOE.

Cortner, H. & Moote, M. (1994). Trends and issues in land and water resource management: Setting the agenda for change. *Environmental Management, 18(2),* 167–174. doi: https://doi.org/10.1007/BF02393759

Daniels, S.E. & Walker, G.B. (1996). Collaborative learning: Improving public deliberation in ecosystem-based management. *Environmental Impact Assessment Review, 16(2),* 71–102. doi: https://doi.org/10.1016/0195-9255(96)00003-0

Eden, S. & Tunstall, S.M. (2006). Ecological versus social restoration? How urban river restoration challenges but also fails to challenge the science-policy nexus in the United Kingdom. *Environment and Planning C: Politics and Space, 24,* 661–680. doi: https://doi.org/10.1068/c0608j

Eden, S., Tunstall, S.M. & Tapsell, S.M. (2000). Translating nature: River restoration as nature-culture. *Environment and Planning D: Society and Space, 18,* 257–273. doi: https://doi.org/10.1177/026377580001800101

Everard, M. & Moggridge, H.L. (2012). Rediscovering the value of urban rivers. *Urban Ecosystems, 15,* 293–314. doi: https://doi.org/10.1007/s11252-011-0174-7

Fabos, J.G. & Ahern, J. (1996). *Greenways: The Beginning of an International Movement.* Amsterdam: Elsevier Science.

Fabos, J.G., Milde, G.T., & Weinmayer, V.M. (1968). *Frederick Law Olmsted, Sr.: Founder of Landscape Architecture in America.* Amherst, MA: The University of Massachusetts Press.

Federal Interagency Stream Restoration Working Group (FISRWG). (1998). *Stream Corridor Restoration: Principles, Processes and Practices*. Washington, DC: U.S. Gov. Printing Office.

Fripp, J., Robinson, K.M. & Bernard, J. (2003). Developing an NRCS stream restoration guide. In M. Clar, D. Carpenter, J. Gracie & L. Slate (Eds.). *Proceedings from ASCE 2003: Protection and Restoration of Urban and Rural Streams* (pp. 143–148). Philadelphia, PA: American Society of Civil Engineers

Gobster, P.H. & Westphal, L.M. 1998. *People and the River: Perception and Use of Chicago Waterways for Recreation*. Milwaukee, WI: National Park Service, Rivers Trails & Conservation Assistance Program.

Gobster, P.H. & Westphal, L.M. (2004). The human dimensions of urban greenways: Planning for recreation and related experiences. *Landscape and Urban Planning, 68(2–3)*, 147–165. doi: https://doi.org/10.1016/S0169-2046(03)00162-2

Gracie, J. & Clar, M. (2003). Issues in stream restoration and protection. In M. Clar, D. Carpenter, J. Gracie & L. Slate (Eds.). *Proceedings from ASCE 2003: Protection and Restoration of Urban and Rural Streams* (pp. 1–14). Philadelphia, PA: American Society of Civil Engineers

Haynes, E.C. (2013). Current of change: An urban and environmental history of the Anacostia River and Near Southeast waterfront in Washington, D.C. (Unpublished senior thesis), Pitzer College, Claremont, CA. Retrieved from http://scholarship.claremont.edu/cgi/viewcontent.cgi?article=1033&context=pitzer_theses

Heritage Conservation and Recreation Service (HCRS). (1980). *Urban Waterfront Revitalization: The Role of Recreation and Heritage*. Washington, DC: USDI, HRCS.

Hopkins, A. (2005). *Groundswell: Stories of Saving Place & Finding Community*. San Francisco, CA: Trust of Public Land.

Keystone Stream Team. (2003). *Guidelines for Natural Stream Channel Design for Pennsylvania Waterways*. Alliance for the Chesapeake Bay [online] www.streamteamok.net/OST%20documents/Nat%20Stream%20Channel%20Design%20Guide%20Penn%202003.pdf

Kondolf, G.M. (1998). Lessons learned from river restoration projects in California. *Aquatic Conservation: Marine & Freshwater Ecosystems, 8*, 39–52. doi: 10.1002/(SICI)1099-0755(199801/02)8:1<39::AID-AQC250>3.0.CO;2-9

Kusler, J. & Larson, L. (1993). Beyond the ARK: A new approach to U.S. floodplain management. *Environment: Science and Policy for Sustainable Development, 35(5)*, 7–11, 31–34. doi: http://dx.doi.org/10.1080/00139157.1993.9929099

Laison, K.L. & Lach, D. (2008). Participants and non-participants of place-based groups: An assessment of attitudes and implications of public participation in water resource management. *Journal of Environmental Management, 88*, 817–830. doi: https://doi.org/10.1016/j.jenvman.2007.04.008

Lave, R. (2012). *Fields and Streams: Stream Restoration, Neoliberalism, and the Future of Environmental Science*. Athens, GA: The University of Georgia Press.

Lave, R., Doyle, M.W. & Robertson, M.M. (2010). Privatizing stream restoration in the U.S. *Social Studies of Science, 40 (5)*, 677–703. doi: https://doi.org/10.1177/0306312710379671

Little, G. E. (1990). *Greenways for America*. Baltimore, MD: Johns Hopkins Press.

Lubell, M., Schneider, M., Scholz, J.T. & Metz, M. (2002). Watershed partnerships and the emergence of collective action situations. *American Journal of Political Science, 46(1)*, 148–163. doi: 10.2307/3088419

McGinnis, M.V. (1999). Making the watershed connection. *Policy Studies Journal,* *27(3)*, 497–501. doi: 10.1111/j.1541-0072.1999.tb01982.x

McGinnis, M.V., Wooley, J. & Gamman, J. (1999). Forum: biological conflict resolution: Rebuilding community in watershed planning and organizing. *Environmental Management 24(1),* 1–12. doi: 10.1007/s002679900210

McDonald A., Lane, S.N., Hancock, N.E. & Chalk, E.A. (2004). River of dreams: On the gulf between theoretical and practical aspects of an upland river restoration. *Transactions of the Institute of British Geographers, 29,* 257–281. doi: 10.1111/ j.0020-2754.2004.00314.x

Moran, S. (2010). Cities, creeks, and erasure: Stream restoration and environmental justice. *Environmental Justice, 3(2),* 61–69. doi: https://doi.org/10.1089/ env.2009.0036

Moran, S. (2007). Stream restoration projects: A critical analysis of urban greening. *Local Environment, 12(2),* 111–128. doi: https://doi.org/10.1080/13549830601133151

Moran, S. (2003). Stream restoration: Opportunities for synthesis and integration. *Journal of Geography, 102*(2l), 67–79. doi: https://doi.org/10.1080/ 00221340308978524

Moran, S., Perreault, M. & Smardon, R. (2013). Finding our way: urban waterway restoration and participation process. In J.G. Fabos, M. Lindhult, R.L. Ryan & M. Jackson (Eds.), *Proceedings of Fabos Conference on Landscape and Greenway Planning 2013: Pathways to Sustainability* (pp. 20–35). Amherst, MA: University of Massachusetts.

National Park Service (NPS). (1996). *Floods, Floodplains and Folks: A Casebook in Managing Rivers for Multiple Uses.* Washington, DC: NPS Rivers, Trails and Conservation Assistance Program.

National Park Service, Association of State Wetland Managers and Association of State Floodplain Managers (NPS/ASWM). (1991). *A Casebook for Managing Rivers for Multiple Uses.* Washington, DC: NPS.

National Research Council (NRC). (1992). *Restoration of Aquatic Ecosystems: Science, Technology and Policy.* Washington, DC: National Academy Press.

Onondaga Environmental Institute (OEI). (2009). *Onondaga Creek Conceptual Revitalization Plan.* Onondaga Environmental Institute, Syracuse, NY. Retrieved from www.oei2.org/OEIResources_OCRPDRAFT.html

Ortiz, B. (1989). Redesign of a flood control project by citizen initiative. In D.L. Abell (Ed.), *Proceedings of the California Riparian Systems Conference 1988: Protection, Management, and Restoration for the 1990s* (pp. 495–500). Berkeley, CA: Pacific Southwest Forest and Range Experiment Station, Forest Service, U.S. Department of Agriculture.

Otto, B., McCormick, K. & Leccese, M. (2004). *Ecological Riverfront Design: Restoring Rivers, Connecting Communities.* Chicago, IL: APA Planning Advisory Service.

Palmer, M.A., Hart, D.D., Allan, D.J., Bernhardt, E., & National Riverine Restoration Science Synthesis Working Group. (2003). Bridging, engineering, ecological and geomorphic science to enhance riverine restoration: Local and national efforts. In M. Clar, D., Carpenter, J. Gracie & L. Slate (Eds.). *Proceedings from ASCE 2003: Protection and Restoration of Urban and Rural Streams* (pp. 29–37). Philadelphia, PA: American Society of Civil Engineers.

Petts, G.E., Heathcote, J. & Martin, D. (Eds.) (2002). *Urban Rivers: Our Inheritance and Future.* London: IWA Publishing and Environment Agency.

Petts, J. (2006). Managing public engagement to optimize learning: Reflections from urban river restoration. *Human Ecology Review, 13(2),* 172–181.

Petts, J. (2007). Learning from learning: Lessons from public engagement and deliberation in urban river restoration. *The Geographical Journal, 173(4),* 300–311. doi: 10.1111/j.1475-4959.2007.00254.x

Platt, R.H. (2006). Urban watershed management sustainability: One stream at a time. *Environment, 48(4),* 26–42. doi: https://doi.org/10.3200/ENVT.48.4.26-42

Platt, R.H. (1986). Floods and man: A geographer's agenda. In R. W. Kates & I. Burton (Eds.) *Geography, Resources and Environment Vol. II Themes of the Work of Gilbert F. White* (pp. 28–68). Chicago, IL: The University of Chicago Press.

Powell, M. (2010). Comment: The Anacostia River: Urbanization, pollution, EPA failures and the collapse of the public trust doctrine. *Law Forum 41(1),* 68–88. Retrieved from https://scholarworks.law.ubalt.edu/cgi/viewcontent.cgi?article=2321&context=lf

Riley, A.L. (1989a). Overcoming federal water policies: The Wildcat-San Pablo Creeks case. *Environment, 31(10),* 12–31. doi: 10.1080/00139157.1989.9928987

Riley, A.L. (1989b). The Wildcat-San Pablo Creek Flood Control Project and its implications for the design of environmentally sensitive flood management plans. In D.L. Abell (Ed.), *Proceedings of the California Riparian Systems Conference 1988: Protection, Management, and Restoration for the 1990s* (pp. 485–490). Berkeley CA: Pacific Southwest Forest and Range Experiment Station, Forest Service, U.S. Department of Agriculture.

Riley, A.L. (1998). *Restoring Streams in Cities: A Guide for Planners, Policymakers and Citizens.* Washington, DC: Island Press.

Riley, A.L. (2002). A Primer on Stream and River Protection for the Regulator and Program Manager. Technical Reference Circular W.D. 02-#1. San Francisco Bay Region: California Regional Quality Control Board.

Rogge, M.E., Davis, K., Maddox, D. & Jackson, M. (2005). Leveraging environmental, social, and economic justice at Chattanooga Creek: A Case Study. *Journal of Community Practice, 13(3):* 33–53. doi: 10.1300/J125v13n03_03

Rosgen, D. (1994). *Applied River Morphology.* Pogosa Springs, CO: Wildland Hydrology.

Schwartz, J.S., Carpenter, D.D., Slate, L.O., Sinha, S., Brenman, K.E., & MacBroom, J.G. (2003). Research needs for improvement of principles and practices in urban stream improvements. In M. Clar, D., Carpenter, J. Gracie & L. Slate (Eds.). *Proceedings from ASCE 2003: Protection and Restoration of Urban and Rural Streams* (pp. 15–28). Philadelphia, PA: American Society of Civil Engineers.

Smardon, R.C. (2009). *Sustaining the World's Wetlands: Setting Policy and Resolving Conflicts.* New York, NY: Springer.

Smardon, R.C. (2006). Heritage values and functions of wetlands in Southern Mexico. *Landscape and Urban Planning, 74(3–4),* 296–312.

Smardon, R.C. (1992). Regulation of environmentally sensitive areas and resources. In R.C. Smardon & J.P. Karp (Eds.), *The Legal Landscape: Guidelines for Regulating Environmental and Aesthetic Quality* (pp. 151–180). New York, NY: Van Nostrand Reinhold.

Smardon R.C., Felleman, J.P. & Senecah, S.L. (1996). *Protecting Floodplain Resources: A Guidebook for Communities.* Washington, DC: Prepared for the Federal Interagency Floodplain Management Task Force, FEMA publication 268 U.S. Gov. Printing Office.

Spirn, A.W. (1984). *The Granite Garden: Urban Nature and Human Design.* Garden City, NY: Basic Books.

Thigpen, J. (2006.) *Stream Processes: A Guide for Living in Harmony with Streams.* Horseheads, NY: Chemung County Soil and Water Conservation District.

Tunstall, S.M., Penning-Rowsell, E.C., Topsell, S.D.M. & Eden, S.E. (2000). River restoration: Public attitudes and explorations. *Journal of Chartered Institute of Water and Environmental Management,* 14, 363–370 doi 10.111/j.1747-6593.2000.tb00274.x

Vileisis, A. (1997). *Discovering the Unknown Landscape.* Washington, DC: Island Press.

Wessels, A.T. (2010). Place-based conservation and urban waterways: Watershed activism in the bottom of the basin. *Natural Resources Journal,* 50, 539–557. Retrieved from http://digitalrepository.unm.edu/nrj/vol50/iss2/14

White, G.F. (1957). A perspective of river basin development. *Law and Contemporary Problems, 22(2),* 157–184. doi: 10.2307/1190252

White, G.F. (1961). The choice of use in resource management. *Natural Resources Journal,* 1, 23–40. Retrieved from http://digitalrepository.unm.edu/nrj/vol1/iss1/2

White, G.F. (1966). Optimal flood damage management: Retrospect and prospect. In A.V. Kneese & S.C. Smith (Eds.), *Water Research; Economic Analysis, Water Management, Evaluation Problems, Water Reallocation, Political and Administrative Problems, Hydrology and Engineering, Research Programs and Needs* [papers] (pp. 251–269). Baltimore, MD: Johns Hopkins Press for Resources for the Future.

White, G.F. (1969). *Strategies of American Water Management.* Ann Arbor, MI: University of Michigan Press.

Wolch, J. (2007). Green urban worlds. *Annals of the Association of American Geographers 97(2),* 373–384. doi 10.1111/j.1467-8306.2007.00543.x

Woolley, J.T., McGinnis, M.V. & Kellner, J. (2002). The California watershed movement: Science and politics of place. *Natural Resources Journal,* 42, 133–183. Retrieved from www.jstor.org/stable/24888820

Wrenn, D.M., Casazza, J. & Smart, E. (1982). *Urban Waterfront Development.* Washington, DC: Urban Land Institute.

2   History of urban river restoration in Europe

Richard Smardon

Introduction: The urban problem

This chapter reviews the recent European history and experience regarding river conservation and restoration with specific reference to urbanized waterways. Basic threats to river corridors will be reviewed as well as European regulations/guidance, the development of the science, the role of environmental NGOs plus the River Skerne and Tame case studies as examples of urbanized water body rehabilitation projects.

In Europe (see Figure 2.1) the major threats to river corridors in general include:

- Increased flooding aggravated by loss of floodplain via development, land raising, waste disposal and agricultural development. Such encroachment is further encouraged by early acts such as the U.K. Land Drainage Act of 1930 and 1946 and the Danish Watercourse Act of 1949 and 1962;
- Increased water runoff within urbanized areas;
- Increased risk of pollution via agricultural intensification, industrialization, sewage effluent and surface water runoff—all of which become more concentrated in low flow conditions; and
- Lower flows in rivers via increased abstraction and concentration of effluent discharges (Gardiner & Cole, 1992, p. 400).

More specific problems with urban river conservation include stream river disturbance from land clearance, construction, paving, lawn maintenance, industrial pollution, human and animal waste and vehicle-related discharges. The alteration of the hydrologic cycle affects stream morphology, water quality and aquatic habitat. Political tensions arise when urban users gain control of water use from rural and agricultural users. Minority groups are also demanding more of a voice as water management decisions affect water quality and quantity (Baer & Pringle, 2000, p. 388).

European literature on river conservation and restoration was generated through the 1990 York Conference (Boon et al., 1992) and subsequent later

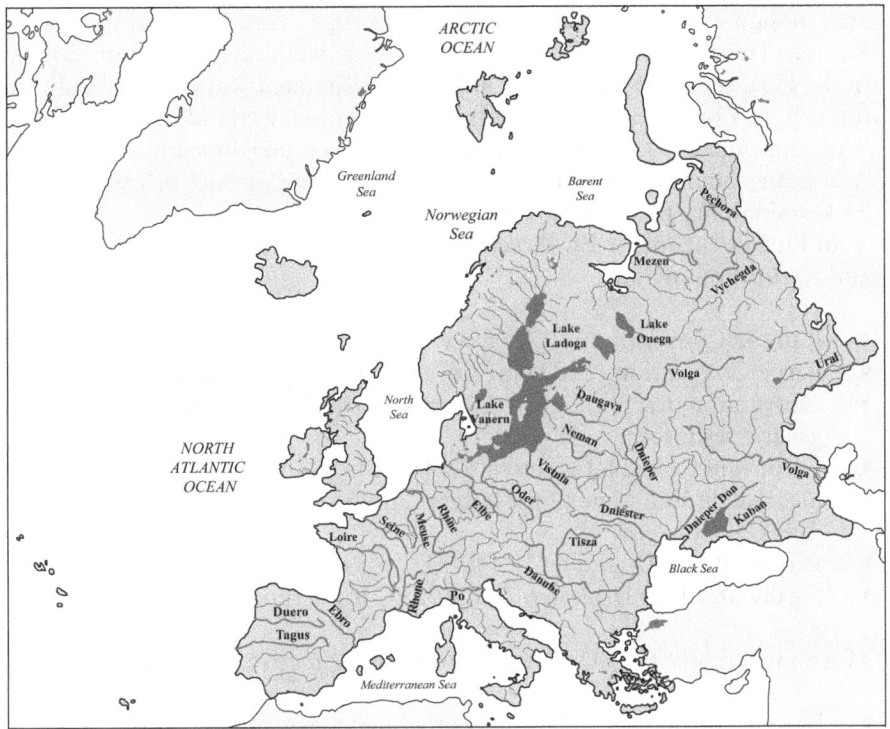

Figure 2.1 European major rivers
Source: Based on Boon et al. (1992), redrawn by Ryan Mackerer

meetings and books (Boon et al., 2000; Boon & Pringle, 2009; Boon & Raven, 2012). Most of this work stresses ecological restoration of nonurban rivers. There is also literature on urban river greenways that stresses cultural and ecological attributes to provide aesthetic, recreational and cultural benefits for urban populations. This work can be found in greenway conferences and proceedings (Fabos & Ahern, 1996; Fabos et al., 2013). Specific examples include Baschak & Brown (1996) describing an ecological framework for the planning, design and management of river corridors; Cengiz et al. (2011) discusses application of such to the Berne River in Turkey; and Tzolova (1996) on the application of a similar assessment system to the Danube River.

EU regulations and guidance

Unlike in the U.S., international agreements are central to European river/waterway planning and restoration. One example is the Convention on

Biodiversity, which has not been signed by the U.S. but is key to protected area designation and management in Europe. Another example is the Ramsar Treaty of 1971 for the protection of wetlands (Smardon, 2009a). In the U.K. there are some 920,000 ha of designated Ramsar wetlands, but the U.S. has only 130,000 ha with a much larger wetland area (Pringle & Withrington, 2009). European communities have also done much more than American communities with biodiversity planning as part of their Agenda 21 sustainability plans (Smardon, 2008).

In Europe the major EU directives affecting river and waterway planning and restoration include:

- Ramsar Convention (1971) for wetland protection.
- Berne Convention (1982) for flora and fauna protection.
- Convention for Biological Diversity (1992) for protected areas and genetic resources.
- European Council Directive 92/43/EEC for natural habitat management, including Annex I for habitat and Annex II for species plus Special Areas of Concern.
- Surface Water Directive 75/440/EEC for drinking water standards; and
- Freshwater Fish Directive 78/659/EEC for certain fish.

Most importantly the EU Water Framework Directive was created to:

- Prevent further deterioration of aquatic ecosystems, protecting and enhancing their status.
- Promote sustainable water use.
- Reduce water pollution to groundwater and surface water; and
- Contribute to mitigating the effects of floods and drought.

The intent of this directive was to "increase the scale of river conservation and connectivity between upstream and downstream reaches—between river channels, riparian zones and floodplains as well as surface of the riverbed and the subsurface" (Boon, 2012, p. 9).

The Urban Wastewater Directive is also significant for urban waterway quality. Until the 1980s phosphorus was not removed in wastewater treatment plants. Throughout Europe there is a requirement for designation of Sensitive Areas at risk from eutrophication. In 2000 the government agency English Nature lobbied with wildlife NGOs to remove phosphorus at some 65 sewage treatment plants to protect 32 sites of special interest as well as 100 parks. So limits were set for phosphorus under the water quality program throughout the U.K. (Pringle & Withrington, 2009, p. 41).

The Water Framework Directive (WFD) was created to monitor surface waters throughout Europe for "good ecological status" by 2015. The WFD directs EU countries to set standards, exemptions, risk assessment and potential costs for water quality protection/improvement for such surface waters.

As part of the WFP process, River Basin Management Plans (RBMPs) were to be initiated for a limited number of river basins. Under EU law the European Community has the power to act against member EU states by withholding payment of development funds. In 2004 there were some 2,531 of these water-quality cases throughout Europe, of which only a third were resolved. In 2005 Belgium, Denmark, France, Germany, Greece, the Netherlands, Luxembourg and Spain all had outstanding cases. Under the Directive on Integrated Pollution Prevention and Control, final warnings went out to Italy, Spain and Greece (Pringle & Withrington, 2009, p. 43).

The other major issue affecting urban waterways in Europe is flood control, or flood "defence" as it is termed in the U.K. After major droughts in 1989 and 1990 public attention in the U.K. focused on water quantity as well as water quality, along with the need for catchment planning (Gardner & Cole, 1992). Europe and the U.K. have since gone through three phases of land drainage, flood defense and flood risk management with different incremental policy development for each phase (Johnson et al., 2005; Tunstall et al., 2004). Penning-Rowsell (1983) describes the early controversies with proposed drainage works for the Yare River Basin in the U.K.

The Water Act of 1989 created a National Rivers Authority (NRA) in the U.K. Its mission was to:

- Minimize environmental impact of its own proposals for flood defense.
- Influence land use control more effectively, as there was a rapid increase in structural controls for flood defense after World War II.
- Assess cost effectiveness of channelization; and
- Promote direct enhancement of river environments so such enhancements could be carried out on their own merits.

In 1996 the NRA merged into the U.K. Environment Agency (EA). Since 2007 the EA has had a wide mandate as both regulator and provision of flood defense, flood management, fisheries, water resources and water quality—all in one Department for Environment, Food & Rural Affairs (Defra).

River conservation and restoration movement

In 1990 the Nature Conservancy sponsored the York Conference on river conservation. This marked the "turning point away from land drainage and hard engineering to more naturalistic approaches" (Newson, 2012, p. 17). Conservation-minded NGOs allied with scientists became a potent movement against ecological destruction, wetland drainage and river channelization. This movement started legislation for river restoration, but such measures were heavily weighted toward nature conservation rather than social benefits because of countryside and wildlife conservation groups' influence. Early river restoration advocates in Europe had little expertise in designing, undertaking and analyzing interdisciplinary research, little experience of developing practical tools from research and little experience with "stakeholder participation" involving those who would be affected by project action.

During the 1980s and 1990s there was much debate about river restoration, rehabilitation, enhancement and mitigation. According to Holmes & Jones (2012), Nadia Johansava gave a presentation on Czech river restoration and inspired the attendees of the 1990 York Conference to establish a River Restoration Centre (RRC).

And so there was an initial river restoration project in 1992 with EU funding. There were three demonstration projects: the Brede River in Demark and the rivers Skerne and Cole in the U.K. With these pilot projects the center was to monitor benefits, motivate and train practitioners, develop partnerships with common goals, establish costs and benefits of restoration, assess public perception and disseminate information. By 2010 it had settled on three core activities: 1) knowledge exchange; 2) advise, assess, promote, facilitate; and 3) support best management practices.

Table 2.1 River restoration centers and networks in Europe in 2009

Country	Center/network	Website
Belgium	University of Antwerp, Ecosystem Management Research Group (ECOBE)	www.uantwerpen.be/en/research-groups/ecobe/
Denmark	Danish Center for River Restoration (Dansk Ctr. for Vandlobsrestaurering-DCVR)	http://www2.dmu.dk/1_Om_DMU/2_tvaer-funk/3_vires/default_en.asp
Finland	Environment Institute (SYKE)	www.environment.fi/
Italy	Italian Centre for River Restoration (Centro Italiano per la Riqualificazione Fuviale—CIFR)	www.cirf.org/en/home-9/
Netherlands	Netherlands Centre For River Studies (NCR)	www.ncr-web.org/
Romania	Romanian River Restoration Ctr.	www.rowater.ro/sites/en/default.aspx
Russia	Russian Research Institute for Integrated Water Management and Protection (RosNIIVH)	www.wrm.ru/about_eng.php
Spain & Portugal	Iberian Centre For River Restoration (Centro Iberico de Restauracion Fluvial—CIREF)	www.cirefluvial.com/en/
United Kingdom	The River Restoration Center (RRC)	www.therrc.co.uk/

Source: Holmes & Jones (2012, p. 288)

The European Centre for River Restoration (ECRR) has operated for more than 20 years, sharing information and funding restoration projects through the LIFE (L'Instrument Financier pour l'Environnement) Foundation. It existed informally from 1995 to 1999, when a meeting in Sikeberg, Denmark, with 55 participants from 22 countries established a formal European River Restoration Network to improve confidence in implementation, increase knowledge, facilitate information exchange world-wide, improve quantification of benefits and encourage natural river resto-ration centers across Europe. Today there are at least nine European Centers (see Table 2.1).

The European Union network of Special Areas of Conservation includes such rivers as the Avon in England, the Tweed in Scotland, the Tywi in Wales and the Upper Ballindera in Northern Ireland (see Figure 2.2).

Figure 2.2 Major United Kingdom rivers
Source: Based on Pringle & Withrington 2009, redrawn by Ryan Mackerer

European NGOs and river restoration

Throughout Europe—especially in northern Europe and the U.K.—environmental NGOs have played a major role in changing the paradigm of river conservation and restoration. This section offers a few examples.

In 1949 the Nature Conservancy (NC) covered all of the U.K. except Northern Ireland. In 1974 the NC split, with research being addressed by the Nature Conservancy Council. In 1979 the NC broke into three organizations: the Scottish National Heritage (SNH), the Countryside Council of Wales (CCW) and English Nature. In 2006 English Nature merged with the Countryside Agency for Landscape and Recreation Future plus the Rural Development Service to become Natural England. (Remember it was the NC that convened the York Conference in 1990 and it was English Nature that lobbied for the Urban Water Quality program to address phosphorus treatment within sewage treatment plants in the U.K.) Other active NGOs in the U.K. include the River Trusts in England and Wales, Rivers and Fishery trusts in Scotland, the Royal Society for the Protection of Birds (RSPB) and the Worldwide Fund for Nature (WWF).

There are anglers' associations in Denmark as well as river keepers and Actin Fishnotterschltz in Germany, plus public awareness/education programs in Norway, Denmark and Germany. But in Northern Europe one of the best examples of collective action is the Trilateral Wadden Sea Plan with Germany, The Netherlands and Demark. This management plan to conserve the Wadden Sea shore, wetlands and river systems came about because of the environmental NGOs in all three countries lobbying for action (Smardon, 2009b).

European case studies

Most river conservation and restoration work in Europe focuses on ecological restoration of waterways versus revitalization. There has been little assessment of social outcomes of river restoration/revitalization projects in Europe and elsewhere (Asakawa et al., 2004; Eden & Tunstall, 2006; Schaich, 2009; Woolsey et al., 2007). Any integrated planning process must encompass social values as well as balance the ecological and social requirements of waterway restoration/revitalization (Gobster et al., 2007; McDonald et al., 2004; Pfadenhauer, 2001).

Case study: River Skerne, Darlington, County Durham, UK

The following is a brief summary of such an assessment for the River Skerne in Darlington, County Durham, UK, and the River Cole near Swindon, Wiltshire, UK, drawn from Aberg & Topsell (2012).

This assessment utilized pre- and post-project surveys in 1997 and 2008 with both structured interviews and survey instruments (with tick box and open-ended questions). The survey instrument asked respondents about the perceived value of the rehabilitated river as well as whether and how perceptions and expectations had changed over time. Data were collected on demographic facts, visit frequency, attractiveness, wildlife, recreation, safety, flooding, public consultation and overall satisfaction (Aberg & Topsell, 2012). Qualitative in-depth interviews were conducted with river recreation professionals.

Results for the urban River Skerne were positive (see Figures 2.3, 2.4 and 2.5), but for the rural River Cole they were mixed. The River Skerne results were positive regarding attractiveness, wildlife, recreation, satisfaction, safety and overall satisfaction. Increased access to the river was key (Aberg & Topsell 2012, p. 250). The local sense of pride, ownership and care was influenced by levels of public involvement. Local residents were consulted about the River Skerne project before construction and had some influence on design features such as footbridges and tree plantings. Schoolchildren were also involved and a local liaison officer acted as mediator/facilitator.

Case study: Upper Tame catchment, West Midlands, City of Birmingham, U.K.

This is the first example of public engagement in the design and implementation of an urban river project with an intensive social learning process (Petts, 2006). This project was funded by the EU Life Program as a Sustainable

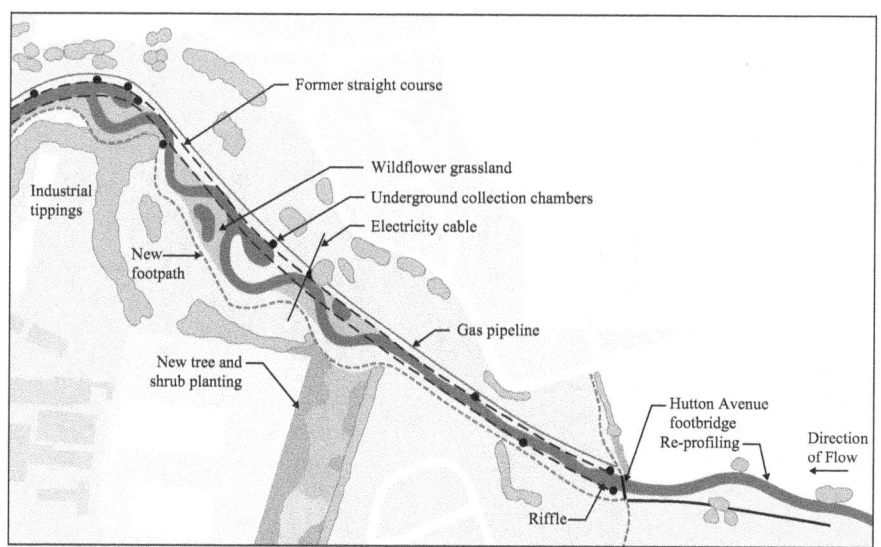

Figure 2.3 Skerne River project

Source: Reproduced with permission from The River Restoration Centre, redrawn by Ryan Mackerer

Figure 2.4 Skerne River backwater marsh
Source: Reproduced with permission from The River Restoration Centre

Figure 2.5 Locomotive footbridge on the Skerne River
Source: Reproduced with permission from The River Restoration Centre

Development of Urban Rivers and Floodplains (SMURF) project. The focus of the SMURF project was regeneration of river corridors and creation of parkland nature reserves. Stretches of the river are in poor ecological condition, have low amenity value and are affected by stormwater flushes, and the river faces development pressures along its floodplain and banks. This was to be an enhancement and/or rehabilitation project as opposed to an ecological restoration.

The comprehensive deliberative social learning process was developed and documented by Petts (2006). Key sub-processes of recruitment of representative interests, active facilitation, collaborative framing of issues, optimizing interaction of experts versus lay people and management of the unexpected are summarized. Key challenges in conducting this process included:

- Recruiting young people under 18 and minority groups.
- Maintaining people's engagement throughout the process.
- Buy-in achieved through capitalizing on local issues that people were concerned about.
- Addressing the tension between what the EU could deliver versus broader community concerns like crime and litter.
- Tension between expert and lay knowledge.
- Provision of appropriate information and management of discussion to promote mutual learning.
- Developing a shared vision tempered by practical constraints and barriers to implementation; and
- Managing the unexpected—what could or could not be dealt with, e.g., security of the playing fields.

According to Petts this two-stage process (stakeholder and local residents) allowed quality time for expert–lay interaction, enabled co-construction of the problem and definition of community priorities and promoted agreement on actions needed and recognition of priority constraints (Petts, 2006, p. 178).

The management plan for the Tame, Anker and Mease Catchment was published in 2013 and includes specific actions for each sub-watershed area (see Figure 2.6).

Pahl-Wostl has also written about the importance of social learning in restoring the multifunctionality of rivers and floodplains in the Netherlands. Use of regional forums, stakeholder platforms in action research and stakeholder processes can all help address uncertainties and ambiguities that arise from stakeholders' different perceptions. Pahl-Wostl proposes group mental model building to address such issues. See Figure 2.7 as an example.

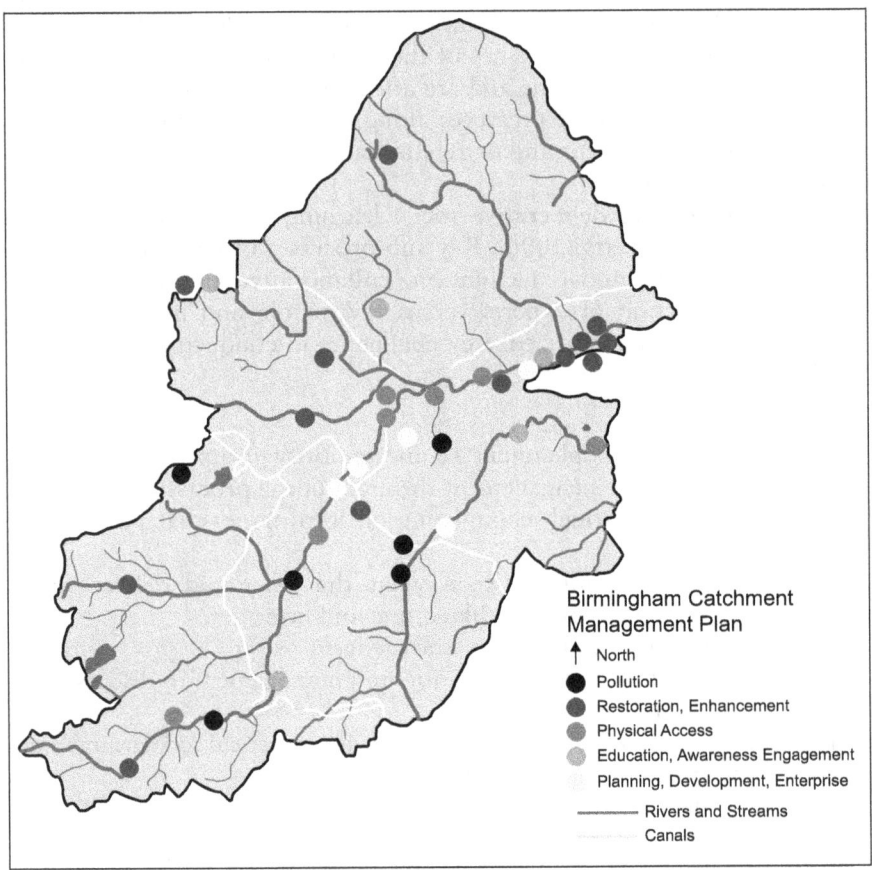

Figure 2.6 Upper Tame–Birmingham Action themes
Source: Based on Environment Agency (2013), redrawn by Ryan Mackerer

Summary

The River Skerne and Tame River case studies illustrate many of the challenges and opportunities of urban river revitalization in Europe. There is a need to incorporate social and ecological service functions with urban waterway restoration projects as well as balance such functions as part of the rehabilitation process. There is also the need to involve stakeholders in fact finding and design aspects of such a process (Moran et al., 2013; Petts, 2006), thereby creating ownership and investment for the stakeholders. Remember in 1990 at the time of the York Conference that there was not much experience with such a multiple-stakeholder process. Lastly there are opportunities to involve stakeholders who traditionally have not been part of such processes, so this could be considered a procedural environmental justice issue.

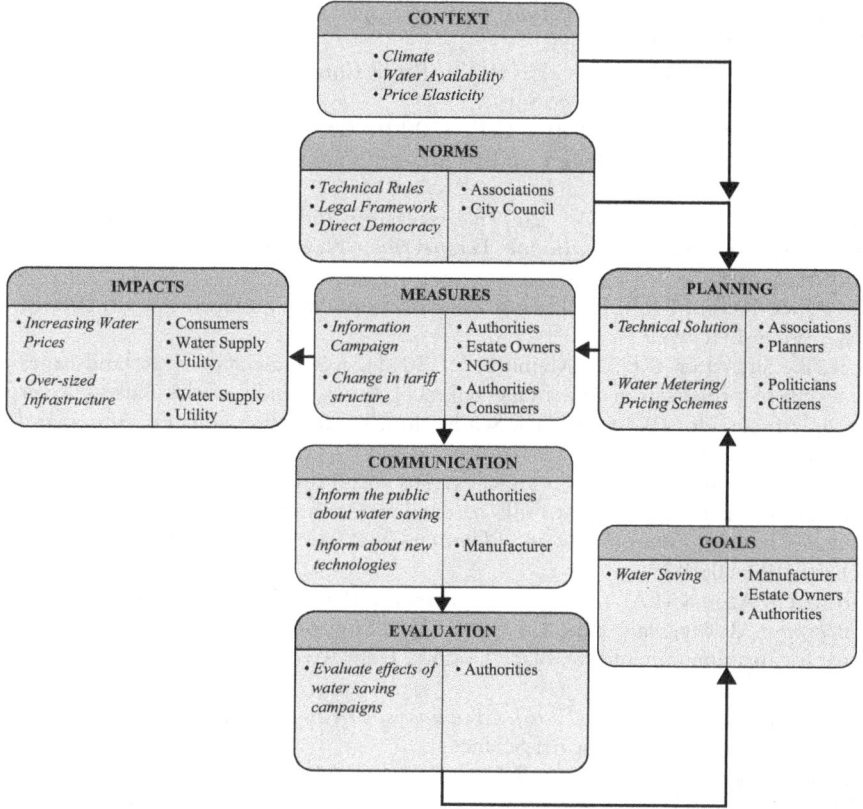

Figure 2.7 Example of a structural model for a water savings scenario
Source: Pahl-Wostl (2006, redrawn by Ryan Mackerer)

References

Aberg, E.U. & Topsell, S. (2012). Rehabilitation of the River Skerne and River Cole, England: A long-term perspective. In P.J. Boon & R. J. Raven (Eds.), *River Conservation and Management* (pp. 249–259). Chichester, U.K.: Wiley-Blackwell.

Asakawa, S., Yoshida, K., & Yabe, K. (2004). Perceptions of urban stream corridors within the greenway system of Sapporo, Japan. *Landscape and Urban Planning*, 68, 167–182. doi: 10.1016/S0169-2046(03)00158-0

Baer, K.E. & Pringle, C.M. (2000). Special problems of urban river conservation: The encroaching megalopolis. In P.J. Boon, B.R. Davis & G.E. Petts (Eds.) *Global Perspectives on River Conservation: Science, Policy and Practice* (pp. 385–402). Chichester, U.K.: John Wiley & Sons.

Baschak, L.A. & Brown, R. D. (1996). An ecological framework for the planning, design and management of urban river greenways. In J.G. Fabos & J. Ahern (Eds.), *Greenways: The Beginning of an International Movement* (pp. 211–226). Amsterdam: Elsevier Science.

Boon, P.J. (2012). Revisiting the case for river conservation. In P.J. Boon & R.J. Raven (eds.) *River Conservation and Management* (pp. 3–14). Chichester, U.K.: Wiley-Blackwell

Boon, P.J., Calow, P. & Petts, G.E. (1992). *River Conservation and Management.* Chichester, U.K.: John Wiley & Sons,

Boon, P.J., Davis, B.R. & Petts, G.E. (2000). *Global Perspectives on River Conservation: Science, Policy and Practice.* Chichester, U.K.: John Wiley & Sons.

Boon, P.J. & Pringle, C.M. (2009). *Assessing the Conservation Value of Fresh Waters: An International Perspective.* New York, NY: Cambridge University Press.

Boon, P.J. & Raven, R.J. (2012). *River Conservation and Management.* Chichester, U.K.: Wiley-Blackwell.

Cengiz, B., Smardon, R.C. & Memluk, Y. (2011). Assessment of river landscapes in terms of preservation and usage balance: A case study of the Barton River floodplain corridor (Western Black Sea Region, Turkey). *Fresenius Environmental Bulletin, 20(7),* 167–1683.

Eden, S. & Tunstall, S. (2006). Ecological versus social restoration? How urban river restoration challenges but also fails to challenge the science-policy nexus in the United Kingdom. *Environment and Planning C: Politics and Space, 24(5),* 661–680. doi: 10.1068/c0608j

Environment Agency (EA). (2013). *Tame, Anker and Mease Catchment Partnership: Catchment Management Plan.* Coventry, U.K.: Environment Agency, The Wildlife Trust for Birmingham and the Black County, Trent Rivers Trust and Warwickshire Wildlife Trust.

Fabos, J.G. & Ahern, J. (Eds.) (1996). *Greenways: The Beginning of an International Movement.* Amsterdam: Elsevier Science.

Fabos, J.G., Lindhult, M.S., Ryan, R.L. & Jackman, M. (Eds.). (2013). *Proceedings of the Fabos Conference on Landscape and Greenway Planning 2013: Pathways to Sustainability.* Amherst, MA: University of Massachusetts. Retrieved from Blogs.umass.edu/faboslgpconf/ (accessed January 17 2017).

Gardiner, J.L. & Cole, L. (1992). Catchment planning, the way forward for river protection in the U.K. In P.J. Boon, P. Calow & G. E. Petts (Eds.) *River Conservation and Management* (pp. 397–406). Chichester, U.K.: John Wiley & Sons.

Gobster, P.H., Nassauer, J.I., Daniel, T.C. & Fry, F. (2007). The shared landscape: What does aesthetics have to do with ecology? *Landscape Ecology* 22:959–72.

Green, C.H. & Tunstall, S.M. (1992). The amenity and environmental value of river corridors in Britain. In P.J. Boon, P. Calow & G.E. Petts (Eds.) *River Conservation and Management* (pp. 425–441). Chichester, U.K.: John Wiley & Sons.

Holmes, N.T.H. & Jones, M. (2012). The history, development, role, and future of river restoration centres. In P.J. Boon & R.J. Raven (Eds.) *River Conservation and Management* (pp. 285–293). Chichester, U.K.: Wiley-Blackwell.

Johnson, C. L., Tunstall, S.M. & Penning-Rowsell, E.C. (2005). Floods as catalysts for policy change: Historical lessons from England and Wales. *Water Resources Development, 21(4),* 561–575.

McDonald, A., Lane, S.N., Heycock, N.E. & Chalk, E.A. (2004). Rivers of Dreams: On the gulf between theoretical and practical aspects of an upland river restoration. *Transactions of the Institute of British Geographers, 29,* 257–281.

Moran, S., Perreault, M. & Smardon, R.C. (2013). Finding our way: Urban waterway restoration and participatory processes. In J.G. Fabos, M.S. Lindhult, R.L. Ryan & M. Jackman (Eds.). *Proceedings of the Fabos Conference on Landscape and Greenway Planning 2013: Pathways to Sustainability* (pp. 20–35). Amherst, MA: University of Massachusetts. Retrieved from Blogs.umass.edu/faboslgpconf (accessed January 17 2017).

Newson, M.D. (2012). From channel to catchment: A 20-year journey in England and Wales. In P.J. Boon & R.J. Raven (eds.) *River Conservation and Management*, Wiley-Blackwell, Chichester, U.K., 17–27.

Pahl-Wostl, C. (2006). The importance of social learning in restoring the functionality of rivers and floodplains. *Ecology and Society 11(1)*, 1–10. Retrieved from www.ecologyandsociety.org/vol11/iss1/art10/ (accessed January 17 2017).

Penning-Rowsell, E.C. (1983). An evolution of wetland policy in England and Wales. In R.C. Smardon (Ed.) *The Future of Wetlands: Assessing Visual-Cultural Values* (pp. 25–39). Totowa, NJ: Allenheld Osmun & Co.

Petts, J. (2006). Managing public engagement to optimize learning: Reflections from urban river restoration. *Human Ecology Review, 13(2)*, 172–181.

Pfadenhauer, J. (2001). Some remarks on the socio-cultural background of restoration ecology. *Restoration Ecology*, 9, 220–229. doi: 0.1046/j.1526-100x.2001.009002220.x

Pringle, C.M. & Withrington, D. (2009). Freshwater conservation in action: Contrasting approaches in the U.S.A. and the U.K. In P.J. Boon & C.M. Pringle (Eds.) *Assessing the Conservation Value of Fresh Waters: An International Perspective* (pp. 39–60). New York, NY: Cambridge University Press.

Schaich, H. (2009). Local residents' perceptions of floodplain restoration measures in Luxembourg's Syr Valley. *Landscape and Urban Planning*, 93, 20–30. doi: 10.1016/j.landurbplan.2009.05.020

Smardon, R.C. (2008). A comparison of local Agenda 21 implementation in North American, European and Indian Cities. *Management of Environmental Quality 19(1)*, 118–137. doi: 10.1108/14777830810840408

Smardon, R.C. (2009a). International wetland policy and management issues. In R.C. Smardon (Ed.), *Sustaining the World's Wetlands: Setting Policy and Resolving Conflicts* (pp. 1–20). New York, NY: Springer.

Smardon, R.C. (2009b). The Wadden Sea Wetlands: A multi-jurisdictional challenge. In R.C. Smardon (Ed.), *Sustaining the World's Wetlands: Setting Policy and Resolving Conflicts* (pp. 21–56). New York, NY: Springer.

Tunstall, S.M., Johnson, C.L. & Penning-Rowsell, E.C. (2004). Flood hazard management in England and Wales: From land drainage to flood risk management. World Congress on Natural Disaster Mitigation, 19–21 February 2004. Retrieved from www.researchgate.net/publication/237378237_Flood_Hazard_Management_in_England_and_Wales_From_Land_Drainage_to_Flood_Risk_Management (accessed January 17 2017).

Tzolova, G.V. (1996). An experiment in greenway analysis and assessment: The Danube River. In J.G. Fabos & J. Ahern (Eds.) *Greenways: The Beginning of an International Movement* (pp. 283–294). Amsterdam: Elsevier Science.

Woolsey, S., Capelli, F., Gonser, T., Hoehn, E., Hostmann, M., Junker, B., Paetzold, A., Roulier, C., Schweizer, S., Tiegs, S.D., Tockner, K., Weber, C. & Peter, A. (2007). A strategy to assess river restoration success. *Freshwater Biology*, 52, 752–69. doi: 10.1111/j.1365-2427.2007.01740.x

Websites

Asian River Restoration Center (ARRC): www.a-rr.net
Australian River Restoration Center (ARRC): https://arrc.com.au
European River Restoration Network: www.ecrr.org
EU Water Resource Framework: http://ec.europa.eu/environment/water/water-framework/info/intro_en.htm
International RiverFoundation (IRR): www.riverfoundation.org.au
River Restoration Centre (RRC): www.therrc.co.uk
River Skerne Project: http://wiki.reformrivers.eu/index.php/River_Skerne_EU-LIFE_project
Skerne River case study: www.therrc.co.uk/projects/skerne_brochure.pdf

3 The big picture

Framing environmental justice, political ecology and stream restoration

April Karen Baptiste and Sharon Moran

Introduction

When streams and rivers get revitalized, does environmental equity suffer? It may be that the inescapable result of improving derelict places in a market-oriented society is upscaling, resulting in compounded inequities. Alternatively, since revitalizing usually means reversing the impacts of 20th-century urban developments, it could be that these projects are an excellent context for reworking unjust urban landscapes. These questions will be explored, drawing on two theoretical frameworks and referring to examples of restoration projects.

The Environmental Protection Agency (EPA) provides the most widely accepted definition of environmental justice (EJ). It is defined as the fair treatment and meaningful involvement of all people regardless of race, color, national origin or income with respect to the development, implementation and enforcement of environmental laws, regulations and policies (EPA, 2017). The goals for EJ include all individuals and communities across the nation enjoying the same degree of protection from environmental and health hazards, and equal access to the decision-making process to have a healthy environment in which to live, learn and work.

Additionally, when defining EJ, many grassroots organizations refer to the Principles of Environmental Justice established in 1991 at the First National People of Color Environmental Leadership Summit (United Church for Christ, 1991). These principles describe the specific rights of those affected, such as compensation, reparations and demands related to public policy, and even go as far as to comment on military occupation of lands.

In this chapter, we explore how two lines of thought—environmental justice and political ecology—can be brought to bear on the whole endeavor of stream restoration, and provide examples of what's been done to date. These examples manifest the potential and pitfalls of revitalizing.

Environmental justice: An overview

History of EJ

The basis for environmental justice (EJ) issues as described by most scholars is the disproportionate levels of pollution and other environmental stressors in minority neighborhoods and workplaces (Bullard, 1990; Bullard & Wright, 1989). Historically, most research has looked at how black and other minority groups, particularly in the U.S., cope with these environmental stressors (Taylor, 2002). In the late 1980s to 1990s, the literature on EJ was scarce to nonexistent. Researchers and activists then began to look at EJ from an equality point of view, starting from the perspective that citizens should all be treated equally (United Church of Christ, 1991). The movement grew out of convergences among the civil rights movement, antiwar and antinuclear movements, women's movements and grassroots organizing around environmental issues (Agyeman et al., 2003; Bryant & Hockman, 2005; Bullard, 1990; Taylor, 2002).

The first landmark case for environmental justice was that of Warren County, North Carolina. Residents of this predominantly African-American county protested both the illegal dumping of PCB-laced toxic soil along the sides of their highways and the permanent storage of the waste in a landfill in their county. The massive protests from residents of Warren County brought to the forefront the disproportionate burdens on communities of color, particularly given that experts confirmed that the siting of the landfill was not determined by scientific factors but rather by race (Walker, 2012). This set precedents for future EJ cases.

In the U.S., the environmental justice movement has grown to serve a wide range of causes across multiple demographic groups and geographical scales. The EJ movement has "sought to redefine environmentalism as much more integrated with the social needs of human populations, in contrast with the more eco-centric environmental movement" (Pellow & Brulle, 2005, p. 5). Given the focus on uneven distribution of burdens and benefits, EJ seeks to bring to the forefront the ways in which societal structures enable uneven power dynamics, which results in environmental burdens placed on some groups more than others. Within U.S. borders, groups that are marginalized by race and income tend to be those most affected by environmental justice issues because of the nation's political, social and economic systems.

Framing of EJ

The environmental justice movement in the U.S. has developed a number of theoretical frames that help assess whether injustices are occurring within the environments of marginalized communities. Bullard (1990) raises issues of marginalized communities being seen as paths of least resistance and hence easy targets for both unwanted land uses and for not receiving environmental benefits. Similarly Pulido (2000) contextualizes the concept of white privilege and the structural forces that have perpetuated a system of privilege and disadvantage for majority and minority ethnic groups. These theoretical concepts are further complemented by the work of Capek (1993), Taylor (2002) and Pellow (2000).

Capek (1993) proposed the Environmental Justice frame. This frame has five different components indicating that environmental injustices occur with the violation of four major rights: the right to accurate information, the right to an unbiased hearing, the right to democratic participation and the right to compensation (Capek, 1993). The fifth component raised in Capek's frame is solidarity, suggesting that building networks between and across groups is crucial for EJ struggles to succeed (Capek, 1993, pp. 9–14).

The first criterion, access to information, asks whether all stakeholders have had equal access to accurate information concerning the injustice. The second criterion, right to a public hearing, takes into consideration whether there was an opportunity to have the injustice discussed in a legal setting. The third criterion, democratic participation, refers to whether all stakeholders were involved in the decision-making process. Scholars often ask the question "Who is at the table?" in reference to who is involved in the discussion. The fourth criterion considers whether and how much monetary compensation has been provided to community members to offset the injustice. The last criterion, solidarity, asks whether solidarity exists in the community where an environmental injustice has been incurred. This five-part theory, though a solid base for evaluating cases, is somewhat flawed in that it does not explicitly lay out a threshold for the extent to which each of the criteria must be met for a case to qualify as an environmental injustice.

Taylor (2002) takes a similar view to Capek on environmental justice theory; however, she describes it as the Environmental Justice Paradigm, taking a goals-based perspective. This frame presents six different components and overlaps some with those of Capek's frame. These components are 1) ecological principles; 2) justice; 3) autonomy; 4) corporate relations; 5) policy, politics and economic processes; and 6) social movement building (Taylor, 2002).Pellow (2000) takes a slightly different approach from the former two frames. His frame, called the Environmental Inequality Formation, includes three different components: the need to redefine environmental inequality as a socio-historical process rather than simply as a discrete event, the need to understand that environmental inequality involves multi-stakeholders, and taking a life-cycle analysis to the struggle (Pellow, 2000). The socio-historic process emphasizes that "environmental inequalities are often subjected to ongoing social constructions by different stakeholders," meaning that it is important to understand the history of an issue to comprehensively understand its current form (Pellow, 2000). Multiple stakeholders can be interpreted to mean specifically that each stakeholder often has different interests, which are complex and at times contradict one another. Life-cycle analysis challenges the scholar and activist to look at the full costs and benefits—the ecology—of production and consumption involved.

In addition to the theoretical framing of EJ, two principles of justice have been embraced in EJ literature, distributive and procedural, with race being a major focus for each principle. Distributive issues deal with impacts of environmental burdens on marginalized communities (Pellow & Brulle, 2005; Schroeder et al., 2008), taking equity into consideration in terms of who bears the environmental costs, particularly those associated with economic development (Zerner, 2000; Newell, 2005). Procedural justice looks at the politics and policies that govern resources (Debbane & Keil, 2004; Pellow, 2000; Schroeder et al., 2008). It also examines the exclusion of individuals by ensuring the absence of meaningful institutional spaces to address the impacts of policies on communities (Newell, 2005). Further it looks at the lack of representation of groups, lack of participation and inadequate access to information (Newell, 2005). In terms of access, this is conceptualized as the ability to benefit from a clean environment (Newell, 2005), but it also deals with allocation of resources and how this leads to inequities within different societal contexts (Debbane & Keil, 2004; Pulido, 2000).

The conceptualization of EJ has been paramount to the expansion of the movement outside U.S. borders. Environmental justice has entered into conversation in the last decade and a half as notable injustices have been brought to international attention in Europe, Latin America, the Caribbean and Asia. The aforementioned frames do remain applicable, along with several new ideas, but often must be reshaped to address regional and national contexts.

Environmental justice: Applications

From U.S. to global perspectives

The approach to EJ in the U.S. context is very different from that of the global context. The former focuses on siting issues and expounds on the race-class debate to a large extent, while the latter has expanded to include issues of access to resources and conceptualizes marginalization more broadly (Newell, 2005). In its global manifestations, marginality takes into consideration a wider range of variables beyond race and class—for example, livelihoods, language and proximity.

The notion of global EJ has been summarized as taking on a human rights frame (Adeola, 2000) and engaging the distribution of environmental goods and bads among human populations worldwide (Dobson, 1998; Newell, 2005). This concept can be broken down into a number of different components, which recapitulate the evolution of environmental justice analyses in the U.S. context. These components include: 1) the unequal exposure of poorer communities to risks, particularly those from economic activity (Newell, 2005; Aygeman et al., 2003; Pellow & Brulle, 2005); 2) unequal access to environmental goods; and 3) marginalization from the political decision-making arenas (Bullard, 1990; Newell, 2005). Schlosberg indicates that there has been a push to globalize EJ as an explanatory discourse, noting that no fewer than 37 countries have had an instance in which an EJ frame has been applied or used to help clarify the problem (2013). There is significant credibility in understanding EJ from a global perspective, particularly when examining the different stakeholders such as multinational corporations that are involved in the traditional economic development that tends to lead to environmental and social inequalities.

Global environmental injustices cannot always be categorized into neat boxes, as can be seen with cases from South Africa, India and Latin America, to name a few (McDonald, 2002; Newell, 2005; Carruthers, 2008). From the beginning of the movement—which can be assumed to be triggered by the events surrounding Bhopal, India, and Pemex, Mexico (Schroeder et al., 2008)—the concept has expanded to include not just distributive and procedural justice components but also issues of access and recognition. There has been increased vigor regarding global environmental injustices due to the deindustrialization of manufacturing regions in the developed world, which has led to relocation of these toxic facilities in the developing world (Schroeder et al., 2008). Further, global EJ concerns can also be seen through the increase in "capital mobility, trade governance, structural adjustment policies, neoliberal projects and displacement of marginalized groups due to war and violence" (Schroeder et al., 2008, p. 548; Baver & Deutsch-Lynch, 2006). In addition, social movement activists have increasingly made alliances across borders, seeking to cultivate relationships with those suffering similar oppressions, and this has been noted in the case of environmental injustice perpetuated by sectors including mining, hydrofracking and petroleum refining, among others.

From its onset, the movement has identified with some of the same ideals and principles as the U.S. EJ movement, its parent theory (Table 3.1). The theories are inherently related in that they both seek to protect individuals from the disproportionate impacts of environmental hazards (Holifield, 2001). Furthermore, justice in both regards appears to speak to how things ought to be through reasoning, ideals, utility, order and a moral obligation (Mohai et al., 2009). The global concept does, however, differ from the United States concept in that it lacks a commonly accepted definition. Schlosberg and Carruthers interpret this as necessary, suggesting that a broad and pluralistic definition of justice must be used because of the diversity of concerns across the global EJ movements (as cited in Schlosberg, 2013, p. 39) (see Table 3.1 for comparison between U.S. and Global EJ).

EJ and stream restoration

While many scholars have "extended formerly U.S.-based environmental justice analysis into international and global spheres, highlighting, for

Table 3.1 Commonalities between U.S. environmental justice and global environmental justice principles

Characteristics	U.S. EJ	Commonalities	Global EJ
Variables of marginality	Race: literal application (African-American/white binary) Socio-economic class	Socio-economic class	Race: figurative application (global north (white) vs. global south (black/brown)) Socio-economic class Livelihoods
Historical context influences	Slavery, Jim Crow, Civil Rights, Environmental movement	Oppression	Labor rights, colonialism, imperialism, apartheid
Themes	Siting of waste sites Cross state concerns	Exploitation of land and people	Mining, transportation of waste, climate change, hydrological issues, water rights
Space/Spatial	Localized scale Urban focused		Cross border interactions Work across geopolitical boundaries

instance, international trade and the global politics of environmental inequality" (Newell, 2005), the contradictory layers of injustice in artisanal gold mining in Ghana (Tschakert, 2009), the politics of solid waste management in Mexico (Moore, 2008) and conflicts over agricultural development in Brazil (Wolford, 2008; Perreault et al., 2012), river restoration has not been extensively examined from an EJ perspective (Hillman, 2006). The relevant literature on river restoration and EJ emerged around the mid-2000s and has been slowly building since then. The main themes that have been seen across the literature are the need for procedural justice, the importance of solidarity, distributive justice and displacement of marginalized groups.

Urban river restoration takes place against a backdrop of "proposing designs that are harmonious with nature and ecological processes" (Honey-Roses, 2008, p. 2070), yet there is recognition that social justice elements are key to restoration (Lave, 2016). Similarly Moran (2007) mentioned that stream restoration, though intended to serve multiple roles, may often privilege a single view of nature and calls for a more comprehensive and equitable way of looking at stream restoration.

One of the prominent themes that has been raised by scholars is the way in which citizens are involved in the river restoration project. Procedural justice is raised in questions of "who made the restoration decisions, who paid, who benefitted" (Honey-Roses, 2008, p. 2070). Honey-Roses' (2008) review of the case studies in the book *Rivertown: Rethinking Urban Rivers* hinted at increasing citizen involvement in the decision-making processes of river restoration, yet it did not give an in-depth analysis of the questions of procedural justice detailing how the involvement was done in the communities. Moran (2007) indicated that a lot of the early movements to push for stream restoration have been initiated by local communities, which were often independent and self-financed. However, a lot more public funds are now available for stream restoration. This shift in financing has the potential to shift the dynamic of who is involved in the decision-making processes, as Moran (2007) stated that differential government spending may re-create unjust relationships in the local landscape.

Within procedural justice the question of who makes the restoration decisions further highlights the issue of whose voices are being heard. Moran (2007) indicates that often urban residents are not included in planning green spaces, though they are the ones who live in the areas where these green spaces are being developed. A case by Metzger & Lendvay (2006) is one of the few studies that shows how EJ is incorporated into environmental decision-making at the community level. In this case the community of Bayview Hunters' Point, San Francisco, California, participated in gathering data that was to be used in the watershed restoration project. Data from community-based projects are more readily accepted than information provided by scientists from outside the community, and community research assistants are recognized as valuable partners in observing long-term changes and identifying pollution sources. Hence, local participation in environmental monitoring is crucial (Metzger & Lendvay, 2006). Such projects can provide access to community feedback, enhance local awareness and build capacity to address environmental challenges (Metzger & Lendvay, 2006). The involvement of community members in the Bayview restoration project had significant benefits for individuals and community members alike. These included people wanting to go on to do studies in environmental fields and people advocating to get signs for fishing advisories; it also built leadership skills among participants. All these benefits help empower members who seek justice.

The scholarship also indicates reasons why procedural justice is difficult to incorporate in river restoration projects. Moran (2007) is one of the few authors who has attempted to operationalize these barriers. Her work suggests that the metrics used for stream restoration do not take into consideration the urban context and by extension the equity concerns. As such, stream miles do not take the level of human interaction into consideration. Essentially, Moran (2007) argues that the current measures used for stream restoration privilege the rural or spaces that are already rich in ecological attributes, which are often the spaces where marginalized communities may not be located. Hence these voices are not included in the decision-making process of stream restoration.

Another reason equity of communities is not considered involves the technical scientific and engineering details that are used to talk about stream restoration. This can be related to the principles of access to information raised by Capek (1993) in the EJ frame. Capek (1993) argued that communities need access to accurate information but that this information should be in a form that is easily understood and readily accessible to the layperson. Moran's (2007) argument is that using scientific and engineering vocabulary to speak about stream restoration excludes a wide range of constituents who should be included in the conversation. A third theme that is raised as a barrier to procedural justice is the way stream restoration efforts are organized. Moran (2007) claims that in many of these projects, though multiple stakeholders are involved, there are no clear lines of responsibility or accountability.

Metzger & Lendvay (2006), in their case on the Bayview river revitalization, found a couple of challenges with community involvement and by extension moving toward procedural justice. First, community members did not feel comfortable engaging in public speaking, which was a part of the process of involvement in the project. This is something that should be addressed if public participation is to be effective. Second, having more elders involved in the project was recommended, as there was a generational gap between those doing the data collection and those receiving the information. As such there needed to be better representation across the spectrum of community members.

Studies increasingly suggest that including citizens at the local level in the management of wetlands and other restoration projects is important for the success of effective management (Davenport et al., 2010; Lave, 2016). They also indicate that residents want to be consulted before restoration projects so they can provide input in the planning process (Davenport et al., 2010). The 1992 Dublin Statement, the bedrock of contemporary water governance, states that

> water development and management should be based on a participatory approach, involving users, planners and policy-makers at all levels. Public participation, in its various formats and strategies, is closely associated with the discourse of environmental governance, for which the involvement of stakeholders can secure greater social commitment, minimize controversy and operational delays, and even create a civil culture.
> (cited in Ioris & Costa, 2009, p. 134)

Here there is the claim that in looking at water governance there needs to be attention to public participation and by extension procedural justice. Indeed, "what is missing in most of the environmental debate today is an acceptance of the transformative role of public participation, not only as an element of improved decision-making, but the cornerstone of active citizenship and environmental justice" (Ioris & Costa, 2009, p. 136).

In a comparative study of river revitalization in the Duwamish Valley in Seattle, the project had two different visions associated with different stakeholders. The government leaders saw the project as one of economic development—that is, jobs—while the grassroots citizens' groups saw the project as one of protecting the surrounding communities' interest

(McKendry & Janos, 2015). These differing interests can lead to conflicts where the stakeholders with the greater power often get the most benefit in having their interests met.

A second theme that was raised was the element of solidarity in river restoration projects. For example, in the case studies presented in *Rivertown*, the chapters on the Los Angeles River, the Anacostia River, the Guadalupe River and the Coalition to Restore Urban Waters all highlight the grassroots nature of the movements to seek restoration, alluding to the importance of solidarity. The L.A. case includes descriptions of the partnerships created among community leaders, academics, students and a research center, while in the Anacostia example, the public–private partnership was highlighted, particularly with respect to fundraising for the project (Honey-Roses, 2008). Some cases highlight the grassroots nature of the restoration movement; for example, the chapter on CRUW (Coalition to Restore Urban Waters) emphasizes the power of the grassroots in making restoration of urban streams visible even if organizations may dissolve over time (Honey-Roses, 2008). Another study indicates that increasing collaboration in management is important because it "increase[s] social learning, build[s] a shared sense of responsibility, resolve[s] natural resource conflicts and pools resources for action implementation" (Wondolleck & Yaffee, 2000, as cited in Davenport et al., 2010, p. 712).

In the Duwamish River case, solidarity was seen among the community members surrounding the restoration project (McKendry & Janos, 2015). Here the community got actively involved in the planning of the cleanup process. The community members' goal was to make the community rather than the planned economic development the center of the conversation. The organization pushed for the community character to be maintained in an effort to avoid gentrification of the community with the new environmental amenities. In the Calumet River case as documented by McKendry & Janos (2015), there was no unified vision of how environmental cleanup connected to jobs being created through the restoration process. This resulted in a fragmented approach to achieving restoration, showing the importance of solidarity in these revitalization projects.

A third theme in the literature on restoration is the question of who benefits from the restoration process. This goes to the heart of distributive justice, which examines how environmental burdens and benefits are dispersed among various stakeholders. In a study about wetland restoration in the Cache River Wetlands in Illinois, Davenport et al. (2010) indicated that when asked about community perspectives on wetlands restoration, respondents mentioned four main issues of importance:

1. Community participation in project planning.
2. Community burdens of the restoration project.
3. Community benefits of the restoration project.
4. Fear and uncertainty around restoration outcomes.

The second and third points deal with distributive justice. The local community at times bears burdens that are not considered in the restoration process. For example, land acquisition for restoration may affect the local government resource base. Additionally there was an allusion to residents being concerned about poverty and meeting those needs rather than protecting biodiversity, which is not considered under restoration projects. Lave (2016) made the point that the unequal exposure to environmental harms through health impacts by avoiding stream restoration in marginalized communities is another burden that must be considered in revitalization projects.

In terms of benefits there were community expectations from the wetland restoration, such as for recreational opportunities and increased revenue from tourism, both of which were not realized in the Davenport et al. (2010) case study. Another study by McKendry & Janos (2015) also highlighted the importance of community benefits from restoration projects. In their study on the Calumet River in Illinois and the Duwamish River in Seattle, though the plans to clean up the rivers were proposed by city officials, the approaches were categorized as technocratic, focusing on removing invasive species and cleaning up industrial slag, sewage and chemicals while also adding recreational amenities. However, these approaches were critiqued by community members who questioned who would benefit from these changes.

A final theme that was raised about river restoration projects, though not widespread, was that of displacement of communities. Greening, under which stream restoration might fall, is being used as a way to develop economies in some industrialized cities (McKendry & Janos, 2015). This phenomenon has been most pronounced in urban cores, and critics of the use of greening as an economic development strategy have noted that such processes are often accompanied by the displacement of poor and working-class people (McKendry & Janos, 2015). This was tied to gentrification as was seen in the Calumet River case. With the proposal for wetland and river restoration there were concerns about gentrification in the Calumet area. The main reason for this fear of gentrification was the fact that the city had not been responsive to requests from low-income residents for environmental goods, yet rushed to provide these greening amenities with the arrival of the middle class to the area (McKendry & Janos, 2015), indicating the notion of displacement of marginalized communities in the face of restoration initiatives.

Political ecology: An overview

Political ecology has been used primarily by geographers and has promise for application to the present subject. Authors writing in this tradition are academics and have varied understandings of what it offers as well as its origins (Perreault et al., 2015; Zimmer & Bassett, 2003). Briefly, political ecology evolved out of a field known as cultural ecology, which was originally put forward to help explain many of the nature–society challenges identified in the developing world starting in the mid-20th century (Gregory et al., 2009). Geographers as well as anthropologists developed cultural ecology, identifying complex interrelationships between the things we consider to be "natural" and the social structures that had been built around them, primarily in non-industrial societies (Gregory et al., 2009). Some of the major emphases of cultural ecology have been the interdependencies among people and ecosystems, especially in the developing world (e.g., Rappaport, 1984; Bayliss-Smith, 1982). Below we discuss four major strengths of political ecology that relate to our topic, and comment on how newer approaches to analysis can illuminate it.

Political ecology frameworks help researchers go beyond managerial approaches by acknowledging that power relations are built into socio-environmental problems and the institutions they are shaped by and embedded in. Political ecology evolved from cultural ecology as the need to unpack power relations and dynamics become more apparent. Different theorists and practitioners understand the field of political ecology in different ways. One approach, articulated by authors Zimmer & Bassett (2003), stresses the unifying qualities of the political ecology framework. They emphasize the importance of the bringing-together that happens within political ecology, across disciplinary boundaries and across the theoretical–applied divide. For example, they subtitled their political ecology book "an integrative approach to geography and environment-development studies." The goals they set out for themselves include "working toward effective policies and understanding key environment-development issues" (Zimmer & Bassett, 2003, p. 3). The authors seek to be theoretically rigorous and also relevant to practitioners, and they seem to have achieved this, since reviewer Nancy Peluso states that their book succeeds in "transcending traditional environmental management and sustainable development paradigms" (Zimmer & Bassett, 2003, back jacket). The authors name four goals and contributions of their book:

- "to engage both the ecological and the political dimensions of environmental issues in a more balanced and integrated way";
- "to expand the geographical range of political ecological studies to include urban and industrial settings";
- to argue for a "more creative consideration of geographical scale"; and
- to engage in "sustained discussion of research methods in political ecology" (Zimmer & Bassett, 2003, pp. 1–2).

A central strength of political ecology is asking analysts to interrogate the process by which things came to be as they are, especially things deemed *natural*. Political ecology insists on paying attention to the not-so-natural dimensions of apparently natural things. Another way of putting it would be that political ecology is ontological in that it interrogates the character of objects, including constructed things (like dams) as well as organic and hybrid things (like grassy riparian buffers), all of which matter in renaturalizing waterways.

In the course of posing such questions, the theoretical approach offered by political ecology can be seen as directing attention away from a main story and toward the edges of it. For example, in the book *Lawn People* author Paul Robbins (2012a) starts with the issues surrounding the complex and contradictory feature of American exurbs, the lawn. He opens the book with the ways that people think and feel about their own lawns, but as this political ecology exploration proceeds, he leads us through understandings of the material flows in and around lawns, and especially how they came to be as they are.

As part of a political ecology approach, Robbins made good on his promise to explore the underlying dynamics of the relationships that are too often assumed to be as they appear at first glance. By unpacking these historical dynamics of materials flows, Robbins was able to highlight new and different facets of the relationships in and around the North American lawn. By exhorting researchers to (re)examine their framing of problems, political ecology has a reflexive quality. This seems to invite analysts to do something that may never come to closure. The tensions around that approach are indicative of exactly what political ecology does; Robbins called political ecology a "trickster," evoking the mythological character whose job is to destabilize and challenge through mocking and magic (Robbins, 2015).

In the literature of nature–society geography, references to political ecology have become so common some acknowledge a sense that perhaps "we are all political ecologists now" (Bridge et al., 2015, p. 4). More seriously, this statement reflects the extent to which the approach has been taken up by so many people working on nature–society topics. Still, debates continue among political ecologists about relative emphasis and methods, among other things (Robbins 2012b). Some of these debates recapitulate the origins of political ecology; still others engage the extent to which political ecology can be (or will be) useful to practitioners and environmental managers (Lave, 2012), given its fairly philosophical orientation.

A second strength of political ecology is drawing attention to the complexities in causal relationships. Many examples of political ecology studies could be cited, but one of the best-known is Blaikie and Brookfield's book *Land Degradation and Society* (1987) on soil and the social processes that undermine its integrity. Their book was published nearly 30 years ago and remains an exemplar of the value of a political ecology framework. Using several case studies, the authors explain how agricultural practices that shape soils are determined by factors far beyond the farmer's choice. Their book skillfully presents the ways in which dynamics of power, control and authority ripple through communities and contribute to the observed situations. Political ecology encourages researchers to acknowledge and attend to the way the rules and practices around natural things have evolved; recognize the contingencies that these arrangements resulted from; and seek out the dynamic nature of the rules and practices, while contemplating the forces that gave rise to change.

A third strength of political ecology lies in interrogating the complexity of socio-environmental dynamics along time scales, in particular heading off simplistic concepts about change by emphasizing the new and unprecedented quality of socio-environmental changes. Authors working in this tradition are pointing out that there is no longer any "nature" outside of human impact, and "the past is no longer a reliable guide to the future, if it ever was" (Braun, 2015, p. 108). With the recent embrace of the notion of the Anthropocene, the character of human impacts is being more fully scrutinized and interpreted. Furthermore, political ecology guides us to inquire about the reasons renaturalizing is needed in the first place. For those setting out to "renaturalize," the job is not so simple: Braun notes "in place of 'baselines' we have, at best, multiple 'basins of attraction'" (2015, p. 107), and this has generated "knowledge controversies" regarding practices such as "rewilding," in which scholars explore "how, by whom and at what scales future natures are to be invented and composed in a world without ecological foundations or certain ecological knowledge" (Braun, 2015, p. 108). For the current case, that would mean asking: In what ways did urbanized watersheds become degraded places, and through what set of socio-ecological dynamics of urbanized places did they come to be as they are?

Finally, the literature emerging in the subfield known as urban political ecology is especially salient for the topic of waterway restoration and environmental justice. The term urban political ecology was adopted to refer to research on ways that political, economic and ecological processes work together. The literature of urban political ecology (UPE) is growing rapidly in connection with urbanization taking place worldwide. UPE "highlights the city as a commodity consumption and political power structuring and shaping socio-spatial flows of power" (Heynen et al., 2006, cited by Barca & Bridge, 2015, p. 370). One challenge noted with urban political ecology is deploying critical social theory to better clarify the problems faced by urban communities. While some studies addressing urban water systems could be called urban political ecology—for example, Gandy's (2003) *Concrete and Clay* and Karvonen's (2011) *Politics of Urban Runoff*—the approach is still fairly new. It has frequently been deployed in connection with overlapping frameworks, notably science studies and/or STS (science, technology and society). STS cultivates observations about the character of a technology and its situation in social systems, especially the extent to which politics and authority are manifest in the technologies themselves. As watersheds become more urbanized, they are rendered more technological through drainage systems, pipes, pumps and other infrastructural elements. For that reason, STS has fruitful overlaps here.

Clearly the political ecology framework can help direct the research trajectory, by problematizing aspects of an issue in richer and deeper ways. As we have shown, political ecology has an inherently provocative quality that places it in major tension with a managerial framework. Much of the work in the realm of nature–society relationships is scholarly, and yet urban waterway revitalization presents immediate problems with ongoing consequences for living people. Political ecology-oriented studies can also encourage practitioners to identify new and different points of intervention. The political ecology perspective affords them opportunities to engage their challenges at different points in the socio-political structures, and in ways they may not have previously imagined. This relates to way(s) that political ecology highlights and recasts the scale and scope of conflict (turning back to a core focus of geographers). With urban waterways, scale challenges are already built in, so theoretical frameworks that accommodate them are especially necessary.

Flowing toward greater justice

The idea of flowing toward greater justice implies a need for a fusing of political ecological thought and environmental justice principles in examining the ways in which communities that are often not the beneficiaries of revitalization efforts can and should be part of these efforts. The combining of these two theoretical schools of thought is not new; a few scholars in recent years have attempted to bring them together.

Holifield (2009) raised the critique of environmental justice research's emphasis on "disengagement from theory and its political focus on liberal conceptions of distributional and procedural justice" (p. 637). He indicated, further, that "urban political ecology has been proposed as an approach that can both contextualize environmental inequalities more productively and provide a basis for a more radical politics of environmental justice" (p. 637). Some scholars suggest that EJ has remained in the realm of empiricist studies and that it does not engage with theory, though recognizing this may be a limited view of the field. It has been proposed that urban political ecology, specifically that characterized as Marxist Urban Political Ecology (Marxist UPE), provides an underlying theoretical frame for application to EJ. Marxist UPE aims to contextualize the production of uneven socio-environmental landscapes within the broader dynamics of capitalist urbanization. It seeks to provide an understanding of the ways in which environmental inequalities result from social power relations, neoliberal forms of capitalist development and class hegemony (Braun, 2005; Keil, 2003; Holifield, 2009).

Heynen (2016) acknowledges the importance of the Marxist UPE view and goes further by postulating the notion of abolition ecology as another avenue whereby political ecology and environmental justice might be fused, explaining,

Abolition ecology represents an approach to studying urban natures more informed by antiracist, postcolonial and indigenous theory. The goal of abolition ecology is to elucidate and extrapolate the interconnected white supremacist and racialized processes that lead to uneven develop[ment] within urban environments.

(Heynen, 2016, p. 839)

Under this approach there is a call to take a deep history of the urban context and/or employ a long-term perspective to understand environmental inequity. The underlying question under the notion of abolition ecology is

how can internalizing the deep historical spatial logics of the "ghetto," the "plantation," the "colony" and the "reservation" push UPE to wrestle with both the racialization of uneven urban environments and also the abolition of white supremacy from the metabolic processes that produce racially uneven urban environments?

(Heynen, 2016, p. 840)

He argues that within the political ecology literature there is opportunity for further grappling with notions of racial capitalism and the ways in which this has shaped the urban context. We expand this by stating that there is a place for understanding how political ecology and EJ work to better explain how river revitalization has taken place in the urban context and the consequences of these revitalization projects.

Heynen (2016) provides an excellent review of the ways in which UPE and EJ have come together, looking at both a North American and non-North American perspective. From the North American perspective, some of the early work examined race and UPE issues of urban forest, parks, food and air pollution. From a non-North American perspective, Heynen (2016) summarizes as follows: "There are rich historical treatments of interconnected power relations shaping urban environments in expansive empirical ways with explosive potential for future development" (p. 841). He goes on to summarize the many empirical studies on the international scale that have in fact stretched the connections between urban political ecology and EJ (Heynen, 2016).

Some of the more recent work linking these two scholarly traditions was seen in the special 2014 issue of *Race, Space, and Nature*. The editors of the issue indicated that the volume put forth "nested arguments about the way that racialization remains a powerful force in contemporary society, contending that intersections with space and nature offer important lessons about the (de)construction of race" (Brahinsky et al., 2014, p. 1135). Others have made a similar case for combining the two traditions. Agyeman & McEntee (2014, p. 217) indicate, "Race, class, and gender (which are the underlying variables of marginality in the environmental justice scholarship), are already established parameters of UPE."

This thinking can be linked all the way back to scholars of the colonial city. Fanon (1963) made it clear that the settlers' town speaks to the notion that colonials were always full of good things (p. 39), while the native town, speaking about those who were colonized, was "starved of bread, of meat, of shoes, of coal, of light" (p. 39). The same theory can be applied to the development of the urban context as highlighted by Heynen (2016), and therefore it is natural to bring together the two fields of political ecology and environmental justice to better understand these dynamics. Ranganathan & Balazs (2015, p. 405) help strengthen these connections further when they suggest,

> While some have argued that the liberal political philosophy underpinning EJ is at odds with the Marxist roots of UPE we find this to be a narrow conception of both literatures, and one that is perhaps more true about their origins than their emerging trends. We thus aim to build on a repertoire of supple "traveling theory" that takes UPE and EJ beyond their respective "home turfs."

River revitalization efforts, particularly within the urban context, have not fully embraced the underpinnings of political ecology or environmental justice, hence the attention being given in this book. Clearly, though, the "flow toward" notion in the section's title is not fully accurate, since it connotes something that happens on its own, and automatically, and everything we're discussing is the result of concerted efforts sustained over time. This is an argument about the importance of institutions and larger-than-the-individual factors in allowing this work to go forward in ways that make sense and improve those communities that are particularly impacted by river revitalization (or its lack thereof).

References

Adeola, F.O. (2000). Cross-national environmental injustice and human rights issues: A review of evidence in the developing world. *American Behavioral Scientist, 43*(4), 686–706.

Agyeman, J. & McEntee, J. (2014). Moving the field of food justice forward through the lens of urban political ecology. *Geography Compass, 8*(3), 211–220. doi: 10.1111/gec3.12122

Agyeman, J., Bullard, R.D. & Evans, B. (Eds.). (2003). *Just Sustainabilities: Development in an Unequal World*. Cambridge, MA: MIT Press

Barca, S. & Bridge, G. (2015). Industrialization and environmental change. In T. Perreault, G. Bridge, & J. McCarthy, J. (Eds.), *The Routledge Handbook of Political Ecology* (pp. 366–377). New York, NY: Routledge.

Baver, S. & Deutsch-Lynch, B. (Eds.). (2006). *Beyond Sand, Sea, and Sun: Caribbean Environmentalisms*. New Brunswick, NJ: Rutgers University Press

Bayliss-Smith, T.P. (1982). *The Ecology of Agricultural Systems*. Cambridge: Cambridge University Press.

Blaikie, P. & Brookfield, H. (1987). *Land Degradation and Society*. London: Methuen and Co.

Brahinsky, R., Sasser, J. & Minkoff-Zern, L-A. (2014). Race, space, and nature: An introduction and critique. *Antipode*, 46(5), 1135–1152. doi: 10.1111/anti.12109

Braun, B. (2005). Environmental issues: Writing a more-than-human urban geography. *Progress in Human Geography*, 29, 635–650. doi: 10.I191/0309132505ph574pr

Braun, B. (2015). From critique to experiment? Rethinking political ecology for the Anthropocene. In T. Perreault, G. Bridge, & J. McCarthy, J. (Eds.), *The Routledge Handbook of Political Ecology* (pp. 102–114). New York, NY: Routledge.

Bryant, B. & Hockman, H. (2005). A brief comparison of the civil rights movement and the environmental justice movement. In D.N. Pellow, & R.J Brulle, (Eds.), *Power, Justice and the Environment: A Critical Appraisal of the Environmental Justice Movement* (pp. 21–36). Cambridge, MA: MIT Press

Bridge, G., McCarthy, J. & Perreault, T. (2015). Editors' introduction. In T. Perreault, G. Bridge, & J. McCarthy, J. (Eds.), *The Routledge Handbook of Political Ecology* (pp. 3–18). New York, NY: Routledge.

Bullard, R.D. & Wright, B.H. (1989). Toxic waste and the African American community. *The Urban League Review*, 13, 67–75.

Bullard, R.D. (1990). *Dumping in Dixie: Race, Class and Environmental Quality*. Boulder, CO: Westview Press.

Capek, S. (1993). The environmental justice frame: A conceptual discussion and an application. *Social Problems*, 40(1), 5–24.

Carruthers, D. (Ed.). (2008). *Environmental Justice in Latin America: Problems, Promise and Practice*. Boston, MA: MIT Press.

Davenport, M.A., Bridges, C.A., Mangun, J.C., Carver, A.D., Williard, K.W.J. & Jones, E.O. (2010). Building local community commitment to wetlands restoration: A case study of the Cache River Wetlands in Southern Illinois, USA. *Environmental Management*, 45, 711–722. doi: 10.1007/s00267-010-9446-x

Debbane, A. & Keil, R. (2004). Multiple disconnections: Environmental justice and urban water in Canada and South Africa. *Space and Polity*, 8(2), 209–225.

Dobson, A. (1998). *Justice and the Environment: Conceptions of Environmental Sustainability and Theories of Distributive Justice*. Oxford: Clarendon Press.

Environmental Protection Agency (EPA). (2017). *Environmental Justice*. Retrieved from www.epa.gov/environmentaljustice

Fanon, F. (1963). *The Wretched of the Earth*. New York, NY: Grove Press.

Gandy, M. (2003). *Concrete and Clay: Reworking Nature in New York City*. Cambridge, MA: MIT Press.

Gregory, D., Johnson, R., Pratt, G., Watts, M. & Whatmore, S. (2009). Political Ecology, in *The Dictionary of Human Geography*, 5th edition. Oxford, U.K.: Blackwell Publishers, Ltd.

Heynen, N. (2016). Urban political ecology II: The abolitionist century. *Progress in Human Geography*, 40(6) 839–845. doi: 10.1177/0309132515617394

Heynen, N., Kaika, M. & Swyngedouw, E. (Eds.). (2006). *In the Nature of Cities: Urban Political Ecology and the Politics of Urban Metabolism*. New York, NY: Taylor & Francis.

Hillman, M. (2006). Situated justice in environmental decision-making: Lessons from river management in Southeastern Australia. *Geoforum*, 37, 695–707. doi: 10.1016/j.geoforum.2005.11.009

Holifield, R. (2001). Defining environmental justice and environmental racism. *Urban Geography*, 22(1), 78–90.

Holifield, R. (2009). Actor-network theory as a critical approach to environmental justice: A case against synthesis with urban political ecology. *Antipode, 41*(4), 637–658. doi: 10.1111/j.1467-8330.2009.00692.x

Honey-Roses, J. (2008). Review: A snapshot of the urban river restoration movement. *Ecology, 89* (7), 2070–2071. Retrieved from www.jstor.org/stable/27650723

Ioris, A.A.R. & Costa, M.A.M. (2009). The challenge to revert to unsustainable trends: Uneven development and water degradation in the Rio de Janeiro metropolitan area. *Sustainability, 1*, 133–160. doi: 10.3390/su1020133

Karvonen, A.A. (2011). *Politics of Urban Runoff: Nature, Technology, and the Sustainable City*. Cambridge, MA: MIT Press.

Keil, R. (2003). Urban political ecology. *Urban Geography, 24*, 723–738.

Lave, R. (2012). Bridging political ecology and STS: A field analysis of the Rosgen Wars. *Annals of the Association of American Geographers, 102*, 2, 366–382. doi: 10.1080/00045608.2011.641884

Lave, R. (2016). Stream restoration and the surprisingly social dynamics of science. *WIREs Water, 3*, 75–81. doi: 10.1002/wat2.1115

McDonald, D. (Ed.). (2002). *Environmental Justice in South Africa*. Athens, OH: Ohio University Press.

McKendry, C. & Janos, N. (2015). Greening the industrial city: Equity, environment, and economic growth in Seattle and Chicago. *International Environmental Agreements, 15*, 45–60. doi: 10.1007/s10784-014-9267-0

Metzger, E.S. & Lendvay, J.M. (2006). Seeking environmental justice through public participation: A community-based water quality assessment in Bayview Hunters Point. *Environmental Practice, 8*, 104–114. doi: 10.1017OS1466046606060133

Mohai, P., Pellow, D. & Roberts, J.T. (2009). Environmental Justice. *Annual Review of Environment and Resources, 34*, 405–430. doi: 10.1146/annurev-environ-082508-094348

Moore, S. (2008). Waste practices and politics: The case of Oaxaca, Mexico. In Carruthers, D. (Ed.), *Environmental Justice in Latin America: Problems, promise, and practice*, (pp. 119–136). Cambridge, MA: MIT Press.

Moran, S. (2007). Stream restoration projects: A critical analysis of urban greening. *Local Environment, 12* (2), 111–128. doi: 10.1080/13549830601133151

Newell, P. (2005). Race, class, and the global politics of environmental inequality. *Global Environmental Politics, 5*, 70–94.

Pellow, D. & Brulle, R. (2005). Power, justice, and the environment: Toward critical environmental justice studies. In D. Pellow, & R. Brulle, (Eds.), *Power, Justice, and the Environment: A Critical Appraisal of the Environmental Justice Movement* (pp. 1–19). Cambridge, MA: MIT Press.

Pellow, D.N. (2000). Environmental inequality formation: Toward a theory of environmental justice. *American Behavioral Scientist, 43*(4), 581–601.

Perreault, T., Bridge, G. & McCarthy, J. (Eds.). (2015). *The Routledge Handbook of Political Ecology*. New York, NY: Routledge.

Perreault, T., Wraight, S. & Perreault, M. (2012). Environmental injustice in the Onondaga Lake waterscape, New York, USA. *Water Alternatives, 5*(2), 485–506.

Pulido, L. (2000). Rethinking environmental racism: White privilege and urban development in Southern California. *Annals of the Association of American Geographers, 90*(1), 12–40. Retrieved from www.jstor.org/stable/1515377

Ranganathan, M. & Balazs, C. (2015). Water marginalization at the urban fringe: Environmental justice and urban political ecology across the North–South divide. *Urban Geography 36*(3), 403–423. doi: 10.1080/02723638.2015.1005414

Rappaport, R.A. (1984). *Pigs for the Ancestors: Ritual in the Ecology of a New Guinea People.* (2nd ed.). Long Grove, IL: Waveland Press.

Robbins, P. (2012a). *Lawn People: How Grasses, Weeds, and Chemicals Make Us Who We Are.* Philadelphia, PA: Temple University Press.

Robbins, P. (2012b). *Political Ecology: A Critical Introduction* (2nd ed.). Malden, MA: Wiley Blackwell.

Robbins, P. (2015). The trickster science. In T. Perreault, G. Bridge, & J. McCarthy, (Eds.). *The Routledge Handbook of Political Ecology* (pp. 89–101). New York, NY: Routledge.

Schlosberg, D. (2013). Theorizing environmental justice: the expanding sphere of a discourse. *Environmental Politics, 22*(1), 37–55. doi: 10.1080/09644016.2013.755387

Schroeder, R., St. Martin, K., Wilson, B., & Sen, D. (2008). Third World Environmental Justice. *Society & Natural Resources, 21*(7), 547–555. doi: 10.1080/08941920802100721

Taylor, D.E. (2002). *Race, Class, Gender and American Environmentalism.* USDA Tech Report. Portland, OR: USDA, Pacific Northwest Research Station.

Tschakert, P. (2009). Digging deep for justice: A radical re-imagination of the artisanal gold mining sector in Ghana. *Antipode, 41*(4), 706–740. doi: 10.1111/j.1467-8330.2009.00695.x

United Church for Christ. (1991). First Principles of Environmental Justice. Retrieved from www.ejnet.org/ej/principles.html

Walker, G. (2012). *Environmental Justice: Concepts, Evidence, and Politics.* New York, NY: Routledge.

Wolford, W. (2008). Environmental justice and agricultural development in the Brazilian Cerrado. In D. Carruthers (Ed.). *Environmental Justice in Latin America: Problems, Promise, and Practice* (pp. 213–238). Cambridge, MA: MIT Press.

Zerner, C. (2000). *People, Plants, and Justice: The Politics of Nature Conservation.* New York, NY: Columbia University Press.

Zimmer, K.S. & Bassett, T.J. (Eds.). (2003). *Political Ecology: An Integrative Approach to Geography and Environment-Development Studies.* New York, NY: The Guilford Press.

4   Environmental justice leadership and intergenerational continuity

April Karen Baptiste

Introduction

One of the significant concerns of environmental justice scholars and activists is that of intergenerational equity and continuity. Intergenerational equity and continuity is defined as the principle that the present generation should pass on to future generations enough natural resources and sufficient environmental quality that they can enjoy at least a comparable quality of life, and inherit a healthy and sustainable environmental heritage. It seeks a fair distribution of the costs and benefits of a long-term environmental policy when costs and benefits are borne by different generations (Park & Allaby, 2017).

In looking at river revitalization and intergenerational equity, two levels of analysis must be done. First, an examination of "what is" must be done. Here we refer to the current ways in which environmental justice principles are incorporated into river revitalization work as a means of ensuring "what can be." The second level of analysis considers which measures are put in place to ensure that future generations are engaged in the revitalization work as a way to ensure perpetuity in the resource.

"What is": How is EJ incorporated into river revitalization?

When examining the concept of intergenerational continuity in river restoration projects, it is important to understand the profile of the organizations involved. Given that socio-demographic variables are central to ideas of environmental justice and that scholars have highlighted the importance of different groups involved in river restoration as indicated in Chapter 3 (Honey-Roses, 2008; Moran, 2007; Davenport et al., 2010; Metzger & Lendvay, 2006), this chapter profiles and assesses the ways in which environmental justice (and by extension issues of intergenerational equity and continuity) are incorporated or not by organizations involved in river revitalization. One of the keys to intergenerational equity and continuity is in understanding the relationships that exist within and among organizations involved in river restoration work. By examining the structure of the organizations that have been involved in stream restoration, including the demographics of both the organizations' leadership and the communities in which the restoration is taking place, plus the ways in which leadership and membership may transcend generations, we are able to understand how environmental justice plays a role in the restoration process. The cases that are used for analysis include:

1. Anacostia River, Washington, D.C.
2. Bronx River, New York.
3. Mill Creek, Philadelphia.
4. Chattanooga Creek, Tennessee.
5. Onondaga Creek, Syracuse.

The profiles of each organization will be based on the following characteristics:

1. Leadership characteristics.
2. Neighborhood demographics for the restoration.
3. Initiation of the revitalization process.
4. Main reasons for initiating the restoration.
5. Strategies used.
6. Outcomes of revitalization.

A summary of the key findings per case is provided in Table 4.1. From the synthesis of the organizations examined in the case studies, multiple themes arose relating to environmental justice and intergenerational equity and continuity. These include distributive justice, procedural justice, strategies used, sustainability and outcomes.

Synthesis

The history of the various rivers highlighted in this chapter provides a sense of what might have promoted the restoration projects and who got involved. One of the central themes across the rivers was a deep historical connection to indigenous people, their original dependence on the river and its surrounding land and the eventual expulsion or dislocation of these groups from their rightful land resources. Indigenous people, the Mohegans for example, used the Bronx River in pre-European settlement times. Europeans were attracted to the area for beavers and then for industrial use, which eventually led to the demise of the river. Over the years of historical degradation a difference in the quality of the Bronx River in the northern part of the city versus the southern parts was observed—hence claims of environmental justice were brought. Similarly for the Anacostia River "the historical processes of plantation agriculture, military industrialization, and discriminatory zoning policies produced social and ecological

Table 4.1 Summary of environmental justice themes across cases

Case study site			Characteristics			
	Organization leadershipNumber of organizations involved, demographics of the management team	Surrounding neighborhood demographics	Who initiated the process?	Reasons for the revitalization?	Strategies used	Outcomes of revitalization
Anacostia River, D.C. Figure 4.2	Two organizations involved (Anacostia Watershed Restoration Partnership [AWRP]; Groundwork DC) Management of AWRP were members of governmental agencies; 60:40 male to female ratio; no indication of the extent of racial diversity; Groundwork DC has a racially diverse leadership and a balanced gender dynamic	Southeast part of the river is predominantly black and lower income Northwest part of the river is predominantly white and wealthy	Joint effort by government, environmental organizations and private citizens	Re-establishing the original ecosystem Focused on health of the ecosystem Vehicle for community development Provide human recreation	Lawsuits, media campaigns	Some displacement of communities
Bronx River, NY Figure 4.1	One main group under two iterations (Bronx River Working Group and Bronx River Alliance) Bronx working group: ethnically, economically and geographical diverse board; seems to be government led Bronx River Alliance: some racial diversity in leadership; more women represented in leadership; seems to be citizen led	Bronx County is over 45.4% non-white; characterized as primarily a black and Hispanic neighborhood One of poorest counties in U.S.	Community activists— initiated in 1974 Since 1997 multiple community organizations, public agencies and businesses have been involved	Restore the river Improve access to the river for the surrounding community	Collaboration between civil society and government Confrontational tactics at times	Achieved significant gains in land along the river Raised public awareness about the river

Chattanooga Creek, TN Figure 4.3	Two main organizations STOP (Stop Toxic Pollution) and APDC (Alton Park Development Corporation)	Alton Park/Pine Wood (AP/PW) is the community most negatively impacted by the pollution; Highly segregated with Alton Park being predominantly African-American and Pinewood being predominantly white; High levels of poverty: 58.1% in Alton Park and 26.2% in Pinewood are below the poverty level	Community groups	Section of creek placed on National Priority List; Heavy contamination from industry; Wanted to address the pollution	Public meetings held by Trust for public land; Charrette held in surrounding communities to determine redevelopment priorities; Meetings held with regional EPA office	Improvement in neighborhood conditions in health services, educational opportunities, housing, job training and abandoned properties
Onondaga Creek, Syracuse, NY Figure 4.5	Two main groups pushed for revitalization; Onondaga Nation Partnership for Onondaga Creek; Government was mainly responsible for implementing the remediation	Onondaga Nation is an indigenous tribal group belonging to the Haudenosaunee; Partnership for Onondaga Creek originated in the Southside community of Syracuse; Mainly an African-American community	Onondaga Nation through land rights lawsuit; County government got involved because of court order (Amended Consent Judgment)	To bring water quality of creek in compliance with regulations	Use gray infrastructure; Regional treatment facility to be built around the city of Syracuse; Lawsuits Onondaga Nation claims treaties led to illegal taking of land; Partnership for Onondaga Creek—filed lawsuit against county to block the RTF from being built	Combination of green and gray infrastructure used along the creek

(continued)

Table 4.1 (Cont.)

Case study site	Characteristics					
	Organization leadershipNumber of organizations involved, demographics of the management team	Surrounding neighborhood demographics	Who initiated the process?	Reasons for the revitalization?	Strategies used	Outcomes of revitalization
Mill Creek, Philadelphia, PA Figure 4.4	The Landscape Literacy Project—Anne Whiston Spirn—white female researcher engaged in community project; research assistants were students from University of Pennsylvania Mill Creek Coalition started in 1999	Mill Creek Neighborhood is predominantly African-American Poorest neighborhood in Philly	Initiated by outside individual and then community	Raise awareness of the environmental hazards of Mill Creek among community members Interested in impact of creek on community including flooded basements, raise awareness of community members	Raise awareness Highlight the indigenous significance of the creek Weekly workshops led by Penn students with middle-school students on history of Mill Creek; summer program with urban gardens	Recognition given to the landscape literacy project; spurred interest to clean up Mill Creek

inequalities in the Near Southeast neighborhood of Washington, D.C."
(Haynes, 2013, p. 6). The Onondaga Nation also historically used the
Onondaga Creek for fishing and recreation. For all of these indigenous
communities, sustenance was interrupted by the beginning of the use of the
river for commercial purposes.

Neighborhood characteristics

From the highlighted cases where we see environmental justice principles
applied there are commonalities in the marginality of the communities that
surround the contaminated river, stream or creek. Demographically, the
South Bronx (Hunts Point, Longwood and Morrisania communities), which
is where a significant amount of the revitalization work is taking place for
the Bronx River, is one of the poorest counties in the country; 43.1 percent
of the population is below the poverty level in this district. In the Hunts
Point and Longwood communities, 55.3 percent of the population is on
income-assistance programs (for the whole borough it is 45.5 percent); while
for Morrisania 61.29 percent of the population is on income-assistance

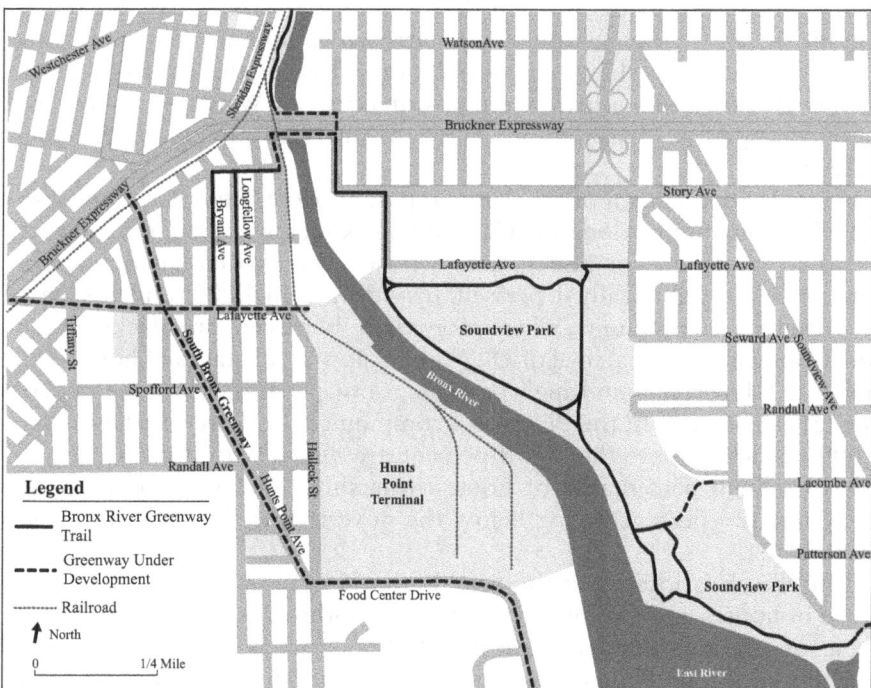

Figure 4.1 South Bronx map
Source: Google, and drawn by Ryan Mackerer

programs (New York City Planning Department, 2015). Additionally, the area has a large racial minority population, where 74.8 percent are of Hispanic origin, 22.1 percent are black non-Hispanic and 1.3 percent are white non-Hispanic.

For the Anacostia River, the District is divided into eight wards, which differ greatly in demographic layout.

> A short car ride from one end of the city to the other reveals significant changes in the quality of living standards, reflecting a move from the majority White Ward in the North of the city to the majority Black Wards in the South. Although it remains a majority African-American city, housing in the District is segregated so that the best environmental quality is clustered away from the city's African-American neighborhoods. Washington's sole majority White Ward boasts the best social services, cleanest water and air, and safest streets.
>
> (Haynes, 2013, p. 6)

The communities most impacted by the river contamination were on the southeast part of the river, which was predominantly black and low income. According to 2010 census data, 35 percent of the District's population is white, and mostly congregate in Wards 2 and 3 in the city's northwest region. In 2010 African-Americans narrowly claimed a majority, rising to 50 percent of Washington's population. Most of the District's northeast, southeast and southwest regions, making up Wards 4, 5, 6, 7 and 8, are occupied by Washington's African-American population (Haynes, 2013, p. 7).

A similar breakdown of the demographics for the Alton Park/Pinewood (AP/PW) community along the Chattanooga Creek was seen, though there seems to be significant segregation between the two communities. This entire area is economically depressed. Just fewer than 4,000 residents live in the Alton Park community, where more than 90 percent are black/African-American with a median age of 28.74 years, low percentages of home ownership (31.1 percent) and high percentages of persons below the poverty line (58.1 percent). In the Pinewood community there are just fewer than 2,000 residents; more than 94.8 percent are white, with a median age of 40 years, high percentages of home ownership (80.4 percent) and relatively few (26 percent) living below the poverty line (U.S. Census, 2010; ACS, 2015).

Mill Creek is located in the neighborhood of West Philly. There are a range of houses and vacant lots in the neighborhood with unkempt houses being one of the dominant features. "Mill Creek is among the poorest neighborhoods in Philadelphia, yet it is home to many well-educated, middle-class residents; almost all are African-American" (Spirn, 2005, pp. 395–396).

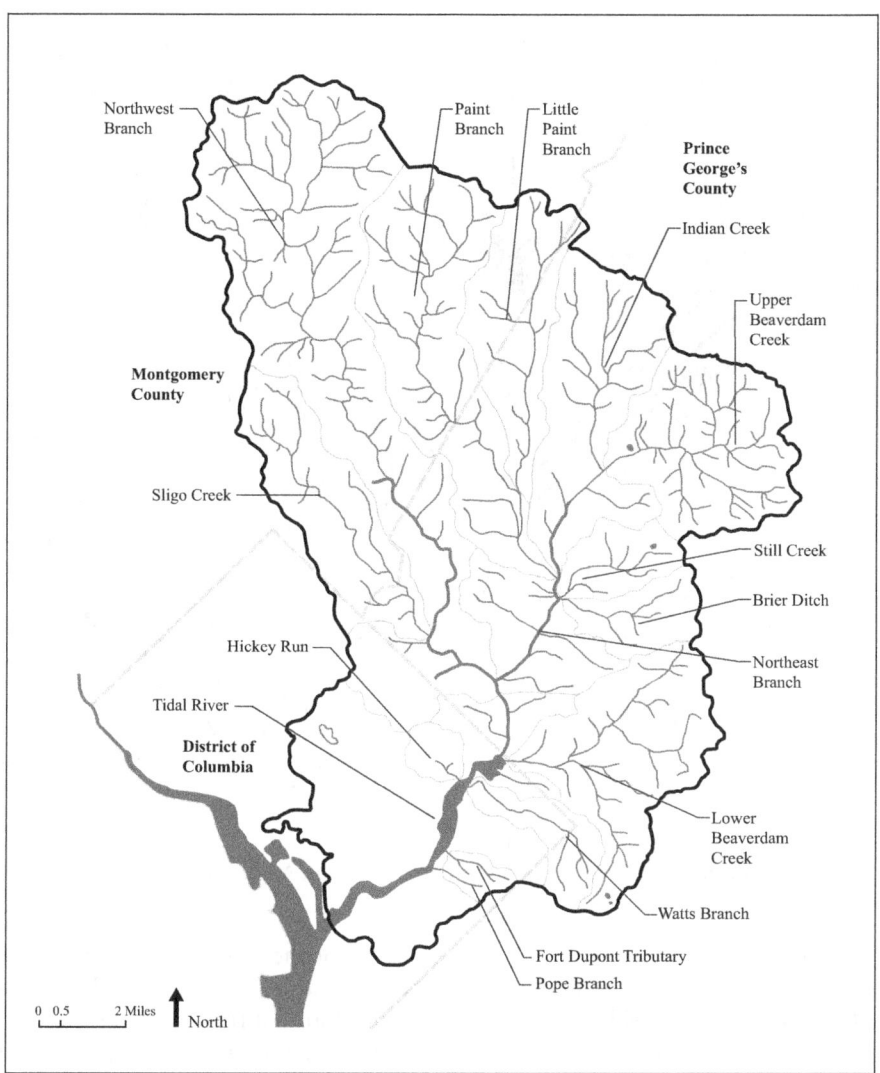

Figure 4.2 Anacostia watershed map
Source: The Summit Fund of Washington and redrawn by Ryan Mackerer

Finally, "along its journey, [the Onondaga] creek connects many communities: farmers, Native Americans, suburbanites, and inner city neighborhoods" (Perreault et al., 2012, pp. 485–486). The two main communities impacted by the contamination of the Onondaga Creek are the Onondaga Nation and the communities of the Southside of Syracuse. Historically, Onondaga Lake

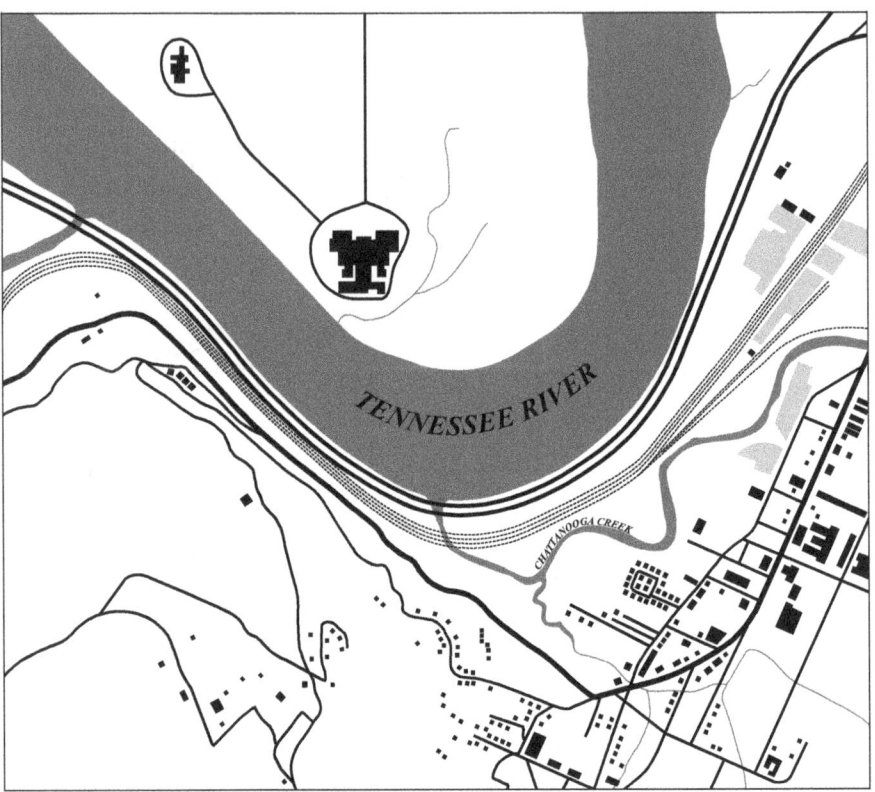

Figure 4.3 Chattanooga Creek map
Source: USGS; redrawn and creek highlighted by Ryan Mackerer

and its watershed formed part of the territory of the Onondaga Nation, one of the six nations of the Haudenosaunee (Perreault et al., 2012). Residents of Syracuse's Southside community are predominantly African-American and low-income, with as few as 15.5 percent to as many as 36.3 percent of the population living below the poverty line in some zip codes (U.S. Census, 2010; ACS, 2015).

The demographics of these aforementioned neighborhoods provide some underlying basis for understanding the injustices that these communities have experienced with river contamination. Within the environmental justice literature it is well established that marginalized communities, i.e., those that are predominantly of racial minority groups and those that are low income, are often targeted for environmental harms (Bullard, 1990; Holifield, 2001; Pellow & Brulle, 2005; Mohai et al., 2009). The historical context of these neighborhoods provides an insight into the distributive injustices they face.

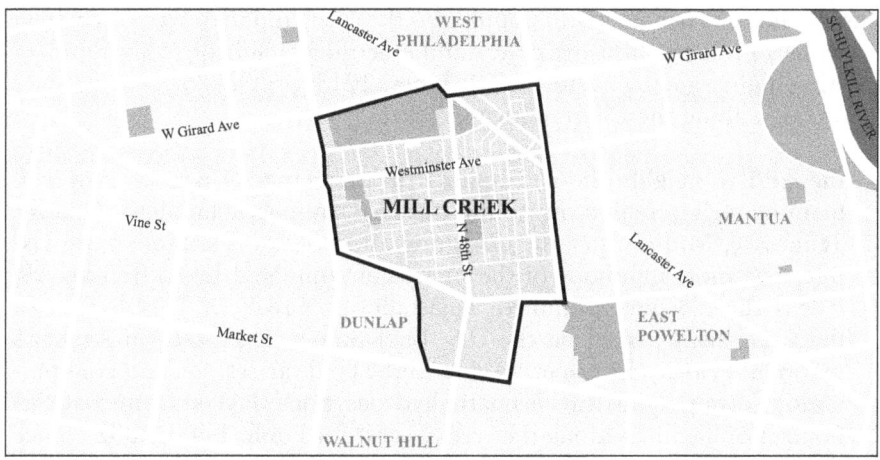

Figure 4.4 Mill Creek neighborhood map
Source: Google and redrawn by Ryan Mackerer

Distributive justice: Who bears burdens and who benefits from the revitalization?

There is a high burden of locally unwanted land uses in the South Bronx area. As Campbell indicated, "Hunts Point and Point Morris are host to more than two dozen waste transfer stations, a sewage treatment plant and a sewage sludge pelletizing plant, along with all of the truck traffic associated with these industrial uses" (2006, p. 39). Given that the area is economically depressed and was inhabited predominantly by people of color and in particular foreign-born persons, an environmental justice ethic was adopted for the cause of the river.

For the Anacostia River case structural inequalities have resulted in the black population experiencing significant levels of pollutants, where the private businesses have focused on the white populations. The neighborhood near the Southeast, which is one of the areas along the Anacostia, has a history of industrial development with pollution being dumped directly into the river (Turner, 2002). This area eventually became an unpleasant environment to live in, resulting in only minority populations residing there.

> As the city grew along racial lines, Near Southeast, with its undesirable environment, was relegated to minority and immigrant residents as the white working class moved across the river to Uniontown (now Anacostia), a rural suburb guarded by racial covenant.
>
> (Haynes, 2013, p. 9)

Today, communities are still exposed to environmental injustices and their consequences, as fishing is still popular on the river, and they are "threatened by changes in public housing policy and emerging trends of redevelopment targeting higher-income residents" (Haynes, 2013, pp. 82–83).

Similarly, along the Chattanooga Creek,

> the AP/PW neighborhood contains the remnants of a once powerful manufacturing region in southern Chattanooga, Hamilton County, Tennessee. Mid-century the city grew and pollution became more visible. The most infamous of the many contaminated sites affecting AP/PW is the 2.5-mile section of Chattanooga Creek ... This section of the Creek was placed on the U.S. Environmental Protection Agency's National Priority List in 1994 due to the coal tar residue and contamination from polynuclear aromatic hydrocarbons (PAHs) from coal carbonization facilities along the creek (Trust for Public Land, 2002). Since the early 1900s, a multitude of industries bordering the Creek, including manufacturing facilities that produced coke, organic chemicals, wood preservatives, and leather products, released contamination into the soil, air and creek sediment. AP/PW residents historically came into regular contact with Creek contamination through fishing and recreating.
> (Rogge et al., 2005, pp. 42–43)

Along Mill Creek, redlining played a significant role in shaping where housing was developed predominantly for African-American citizens, and the history of industrialization also contributed to the demise of the creek.

> By the late 19th century, the creek was polluted by wastes from slaughterhouses, tanneries and households. In the 1880s, it was buried in a sewer, its floodplain filled in and built upon, but it still drains the stormwater and carries all the wastes from half of West Philadelphia and from suburbs far upstream.
> (Spirn, 2005, p. 398)

The Onondaga Creek, Lake, and its watershed were also impacted by industrial production. "Salt production, agriculture, and other development spurred widespread deforestation within the watershed, which would continue through the early 1930s" (Perreault et al., 2012, p. 492). As the population began to expand, the tributaries of the Onondaga Creek and the creek itself began to be used for sewage disposal, while the creek was also channelized and the mouth shifted to avoid flooding. These physical alterations negatively impacted the "Onondaga Nation's traditional fishing, hunting and gathering practices" (Perreault et al., 2012, p. 495). With sewage going into the creek, communities on the Southside of Syracuse were the most impacted by decisions to address the degradation of the Onondaga Creek. It is evident from both the neighborhood characteristics and the historical contexts of these cases (Ducre, 2012) that there are commonalities of marginality associated with river degradation. Part of the deeper interest for this chapter is the ways in which environmental justice principles may or may not be taken into consideration in the revitalization processes of these contaminated waterways.

Procedural justice: Which groups are involved?

An important element of environmental justice is ensuring that all groups regardless of their differences are able to participate in the decision-making process. When looking at river restoration and revitalization projects the question of who is involved in the restoration work is important. Variables such as race, gender, class, educational background and other socio-demographics can be used to assess who is involved in restoration movements, generating postulations as to what the leadership of this movement means for environmental justice work. By this we mean that it is important for the leadership of restoration movements to be representative of the beneficiaries of the revitalization work. For environmental justice, racial diversity is important, particularly as those communities who tend to experience environmental bads are often those who are from underrepresented groups (Bullard, 1990; Pellow & Brulle, 2005). Interestingly for the highlighted cases in this chapter, we see environmental justice being pursued by groups that are demographically representative of the constituents that are impacted by the contamination of the rivers, with the Bronx River case providing the most in-depth example of how environmental procedural justice principles were incorporated into representation.

One group under two different iterations (Bronx River Working Group and Bronx River Alliance—current version) was responsible for spearheading the revitalization effort along the Bronx River.

> Catalyzed by the New York City Department of Parks and Recreation and National Parks Service Rivers and Trails Program, over 60 grassroots groups (including environmental justice groups and non-environmental community-based groups) were brought together, starting in 1996, as the Bronx River Working Group.
>
> (Campbell, 2006, pp. 35–36)

The board of the Alliance consisted of "a number of community and environmental activists, drawn from the participants in the Working Group, and is not just comprised of powerful elites" (Campbell, 2006, p. 47). Members of local environmental and youth-based organizations sit on the board, but so do large, well-resourced environmental nonprofits—for example, the New York Restoration Project and the Wildlife Conservation Society. This "decision was an overt ideological and strategic choice of the group to orient itself more towards its base of community power than to specific elites" (Campbell, 2006, p. 47). Further the Bronx River revitalization effort leadership included a person who was well educated, was local to the area and had urban planning background and experience in navigating foundations. The latter was key for getting access to additional grants and financial resources. Thus in both iterations of the Bronx group the board and leadership were gender-balanced and ethnically, economically and geographically diverse, which was a deliberate attempt to represent the population experiencing the contamination.

Within the Working Group and the Alliance, the role of policy entrepreneurs was helpful in that "the leadership of the Working Group was able to achieve the delicate balance of accessing federal support without allowing it to overwhelm or co-opt community priorities." What this meant was that

> these individuals served as leaders who could help to redefine the traditional roles of their parent agencies, showing that managing the urban environment will require creativity and flexibility on the part of natural resource managers and other types of government agencies.
>
> (Campbell, 2006, p. 42)

Having a relatively large and well-staffed organization (19 full-time staff persons) furthered the work of river revitalization and increased the capacity for community engagement in the revitalization work of the Bronx River.

Similarly, in the cases of the Anacostia River, Chattanooga River and Onondaga Creek, we see the organizations pushing for revitalization attempting to include those communities most severely impacted by the contamination in leadership roles. For example,

> local environmental activists partnered with leaders in the majority African American neighborhoods along the Anacostia waterfront to successfully pressure major polluters like the U.S. Navy to begin a process of meaningful river and waterfront cleanup, and adopt a policy of environmental sensitivity.
>
> (Haynes, 2013, p. 9)

In Chattanooga, two community groups were active in pushing for revitalization of the creek: Stop Toxic Pollution and the Alton Park Development Corporation. The revitalization of Onondaga Creek began with a lawsuit filed by the Onondaga Nation against the county for violation of treaties, which indicated that the land was to be protected to ensure that both communities, i.e., the indigenous Onondaga Nation and the descendants of European colonists, were able to use the resources to support sustenance. The Onondaga Nation claimed that with the contamination of Onondaga Creek and Lake, their sustenance was severely impacted. The Partnership for Onondaga Creek is a community-based organization composed of activists mainly from the Southside of Syracuse who also joined the fight for the revitalization of the creek and lake. The Partnership for Onondaga Creek was created to oppose the building of a sewage treatment facility in a predominantly black community, which was the initial proposal for attempting to clean up Onondaga Creek. The organization's role is complicated in that its leaders wanted the creek cleaned up but not with the solution that was being proposed by the city and county; they advocated alternative methods that did not promote the continued marginalization of the Southside community.

It is only in the Mill Creek case where we see a diversion where the initial person driving for revitalization was an outsider to the community—a professor who at the time was based at the University of Pennsylvania and who was white. However, in this case the researcher collaborated with a local middle school located along the creek and launched an environmental education program that pushed children in the community to begin asking questions regarding the creek. This indirect pressure eventually led to the talks about cleaning up Mill Creek (Spirn, 2005). Given that the representation of the community, the leadership composition and structure of organizations pushing for revitalization work is important, examining who is driving the revitalization projects in terms of initiation and strategies used also plays an important role in advancing environmental justice principles.

Who initiated the revitalization and strategies used?

In almost all of the cases these revitalization projects in marginalized communities were initiated by community members either individually or through the creation of collective groups as described in the previous section. Following the initiation by community groups, government often pitched in. For example,

> in 1999, the Mill Creek Coalition, a group of neighborhood organizations, invited [Spirn] to work with them on the creek and its impact on the community, including research on flooded basements and a course for residents on the history of Mill Creek's landscape.
>
> (Spirn, 2005, p. 407)

The government then got involved through legislation for urban redevelopment. The Onondaga Creek case is another example of the community initiating the revitalization with government getting involved after. The Atlantic States Legal Fund initiated the lake cleanup with legal action in 1988. This resulted in the courts ordering the county to upgrade sewage and combined sewer overflows polluting the lake and its tributaries under the Clean Water Act. The Onondaga Nation later initiated a separate stewardship lawsuit against New York State, which stated that they wanted the lake put back to its condition before European settlement, as a way to recognize the aboriginal territory. Meanwhile the U.S. EPA's declaration of the lake bottom and surrounding areas as a Superfund site led to a remediation effort to clean up the industrial pollution legacy.

This is telling for two reasons: First, it highlights the lack of attention that is given to marginalized communities. Here the claim is being made that even though for decades the rivers and creeks that ran through black, brown and indigenous communities were used as waste dumps, those in power, often white and wealthy, did not see the need to address the contamination. These "black bottoms" (Spirn, 2005) were out of sight and hence out of mind for those in power. Second, this speaks to the notion of who has the efficacy to ask for change. In these cases, community groups sought to push for their voices to be heard in order to improve the living standards in their spaces.

The strategies used by community groups in these cases also reflect environmental justice principles. The Bronx case showed a combination of top-down and bottom-up strategies. The Bronx River Alliance (BRA) at times took a confrontational approach to the revitalization project, engaging in protests or calling media conferences and feeding information to the press. This was seen when the city tried to take over a cement plant to give it to a new industrial site, or when the BRA promoted the decommissioning of the Sheridan Expressway (Campbell, 2006). Confrontational tactics were also used in some of the other cases. For the Anacostia River,

> successful legal battles and media campaigns have led to achievements such as holding the designation of the Washington Navy Yard as an EPA Superfund site (Turner, 2002) and establishing the D.C. bag tax, a five-cent tax on plastic bags that contributes to a clean up fund for the Anacostia.
>
> (Haynes, 2013, pp. 58–59)

The Onondaga Creek revitalization effort was also marred by confrontational tactics used by both the Onondaga Nation and residents of Southside Syracuse.

> On 11 March 2005, the Onondaga Nation filed a lawsuit in U.S. federal court asking for a declaratory judgment that New York State's acquisition of Onondaga lands between 1788 and 1822 was illegal, and that the Onondagas therefore still hold title to approximately 10,360 km² stretching from the St. Lawrence river and the eastern shore of Lake Ontario south to the Pennsylvania border.
>
> (Perreault et al., 2012, p. 496)

In 2010 the Onondaga Nation published a document outlining specific goals for the lake informed by the unique cultural perspectives and needs of its people. The Onondagas were also participating in government-to-government discussions about future lake restoration efforts as part of the Natural Resource Damage Assessment (NRDA) and Restoration process for the Onondaga Lake Superfund site (Perreault et al., 2012), but later withdrew from the NRDA compensation process when they were asked to economically quantify their cultural damage.

> The degradation of the creek's and lake's water quality resulting from municipal waste disposal led to the filing in 1988 of a lawsuit by the environmental organization Atlantic States Legal Foundation (ASLF) and New York State against Onondaga County, which owns and operates the sewer lines fitted with CSOs as well as Metro, the sewage treatment plant that serves Syracuse and surrounding areas.
>
> (Perreault et al., 2012, p. 497)

The county responded by indicating that it would clean up the creek using traditional gray infrastructure, that is, using a regional treatment facility, which is basically a sewage treatment plant to be located in the Southside community. Residents of Syracuse Southside, in forming the Partnership for Onondaga Creek (POC), also filed a lawsuit against the county claiming that their community was targeted for environmental racism. During the construction, POC members' protests included acts of civil disobedience and the filing of a Title VI administrative complaint asking the EPA to rule that the project was discriminatory (Perreault et al., 2012, p. 498).

In some of the cases, though, non-confrontational strategies were used. The Chattanooga Creek community engaged in public meetings that were held by the Trust for Public Land, charrettes were held in surrounding communities to determine redevelopment priorities and meetings held with regional EPA offices. In Mill Creek an environmental education approach was taken as an indirect way to pressure the city to begin cleanup on the creek. Working with the Sulzberger middle school, Spirn states,

> The Mill Creek Project continued … with a format of weekly workshops led by Penn students during the academic year and a four-week summer programme in July based at the school and at Aspen Farms. In 1998, the Sulzberger principal and teachers decided to expand the programme, wrote grant proposals, and obtained funds to do so.
>
> (Spirn, 2005, p. 406)

Whether confrontational or non-confrontational measures were used, in all of these cases bottom-up community mobilization was used to push for the cleanup of the river for the betterment of the community. This worked along with top-down coordination and investment from management agencies and elected officials. These coordination efforts highlight the significance of solidarity in effecting change in river revitalization.

Solidarity

Taylor (2002) and Capek (1993) both indicate that in framing environmental justice, solidarity and social networks are important. These allow groups to find commonalities in their struggles as well as share resources with one another to resist environmental negatives. In all the cases examined, a relatively small number of organizations were listed as being involved in an integral manner in the river revitalization projects, yet it was clear that these key organizations built strong collaborations with other community groups, private organizations and government agencies to advance the goals of revitalization.

Campbell (2006, p. 1) indicated "cooperation is the strategy that can serve to bring the most resources to bear for long term planning and revitalization." For the Bronx River Case, solidarity worked due to internal and external factors. As Campbell (2006, p. 33) indicated,

> The Bronx River partners successfully built a coalition of over 65 organizations, comprised largely of small community based groups and nonprofits matched with a local municipal partner—the Parks Department—and several federal agencies, including the National Parks Service. The fertile civil society provided a grassroots base that was coupled with strategic political alliances through the work of policy entrepreneurs. The coalition was cemented by an environmental justice vision for the future of the river with a flexible organizational structure that incorporated ecological science, open space development, education, and stewardship.

Similarly in the AP/PW community in Chattanooga,

> a strong neighborhood identity and cohesiveness are crucial components that lead to the formation of grassroots collective action at the local level. This interconnectedness is necessary to achieve the visibility and political strength necessary to garner resources and begin the redevelopment process.
>
> (Rogge et al., 2005, p. 36)

The community framed the case as one concerning the neighborhood as a whole seeking to set the goal of taking a holistic approach to the problem.

> While there is no formal political alliance between the POC and the Onondaga Nation, the two groups frequently support each other in pursuing environmental and economic justice. The Onondaga Nation actively supported the POC in filing its Title VI claim with the EPA by providing legal assistance and paying for the printing of necessary documents. Onondaga Nation lawyers assisted the POC to incorporate legally, and have provided advice and advocacy at key moments in its struggle. Indeed, the POC and Onondaga Nation share a common network of support and

solidarity, including legal counsel, a local environmental non-profit organization, and community peace and justice activists.

(Perreault et al., 2012, p. 498)

Solidarity works in revitalization efforts for a number of reasons. First,
the importance of having clear goals for the coalition cannot be overstated
as it allows the groups to have direction, as was seen in all these cases. In
their analysis of the civil rights movement and the environmental justice
movement, Bryant & Hockman (2005) indicated the importance of setting
goals, drawing conclusions that the civil rights movement was able to progress a bit further given that it had a few key goals that were driving the
movement at the time.

A second factor that may contribute to solidarity leading to successful revi-
talization projects is having split responsibilities. For example in the Bronx
River case the sole responsibility did not rest on one entity (Campbell, 2006).
As such, providing multiple opportunities for collaboration and creating an
egalitarian ethos among the groups involved aligns with the ideology of the
EJ movement not having one leader (Bryant & Hockman, 2005).

Third, the availability of shared resources allowed these coalitions to
advance further. "The availability of tangible and intangible resources to a
social movement often determines whether the cause will move forward"
(Rogge et al., 2005, p. 37). The sharing of resources was seen in a couple of
the cases through public–private partnerships. In the Bronx case, where there
were bureaucrats who acted as policy entrepreneurs, coalition building was
considered successful. That is, it is important to engage with persons who are
well aware of the policy pieces needed for the revitalization work but who
are also familiar with getting grants, etc., to support the work. These public–
private partnerships work as by "attracting cooperative and committed political allies at multiple governmental levels, the need for contentious strategies
and pressure-tactics is reduced" (Campbell, 2006, p. 41). Rogge et al. (2005,
p. 36) continue, "Community-based organizations have been shown to achieve
their goals more readily if they collaborate with private industry rather than
competing or attempting to duplicate the role of the private sector."

Finally, the political climate plays a role in determining whether solidarity
can impact revitalization work. In the Chattanooga case, there was a climate
where the downtown area was receiving resources to promote the notion
of an environmental city, yet resources were not being poured into the AP/
PW community, though it was among the most contaminated areas along
the river. This juxtaposed position of the contaminated AP/PW neighborhood against the model Chattanooga downtown environmental city put
pressure on officials to create a somewhat even playing field and hence led
to momentum for revitalization work. Having a climate that is open to civic
resistance allows solidarity to work best and hence advances revitalization
work. The existence of vibrant political party politics in the area where
the restoration was taking place was one of the main reasons the Bronx

River Working Group and the Bronx River Alliance were successful. The Democratic Party machine targets the Bronx, so the area was ripe with political officials to pay attention to interests with strong voices like the Bronx River Working Group. Environmental injustices tend to be easier to address when there is an open and receptive political system. There was an active civic culture with community groups in the Bronx, engaging in revitalization of their blocks, promoting art and cultural events and programs. Hence when the Bronx River Working Group became involved in the community it was welcomed as the community was used to this type of civic engagement.

Sustainability—a balance needed

One of the ways environmental justice principles were seen across the cases was in the balance of traditional environmental goals and environmental justice goals. The former focuses on the importance of the natural resource, its intrinsic values and its protection outside the context of community needs, while the latter focuses on community needs and the ways in which protection of the natural resource enhances overall community well-being. In some of the cases we see that there were splits at times in terms of what the primary focus of the revitalization should be, which caused conflicts.

For example, in the Anacostia River case, government, NGOs and communities focused on ecological restoration while private investors focused on lucrative waterfront development as the cleanup took place (Brendes, 2007; Turner, 2002). "Throughout the debate on city council, environmental concerns about the dangers of building on the Anacostia's ecologically fragile waterfront were markedly absent, and economic worries over the use of public funds consistently trumped environmental issues" (Haynes, 2013, p. 89.) This resulted in environmental concerns and affordable housing playing less of a role than economic development in the cleanup of the Anacostia. A similar situation was seen with Chattanooga Creek, where economic development was seen as the main goal for the revitalization process. As Rogge et al. (2005, p. 49) stated, "The motives of leaders, whose causes ranged from cleanup of pollution and reduction of industrial emissions to the maintenance and improvement of the community's economic infrastructure and housing, included altruism, cultural ties, political advantage, and economic survival." In both of these cases, the ecological importance of the rivers was not balanced with the need for community development.

The Bronx River case shows how sustainability and ecological principles can be incorporated into river restoration without trampling on community interests: First, by seeing the scientific and technical community as allies and not project leaders, which allowed community interests to be at the center of the project while avoiding neglect of the importance of ecological protection. Second, by forcing the scientific community to explain the materials in a way the local community could understand, in plain English, community members were able to understand how community goals work hand in hand with ecological protection. The scientific and technical information served as a "unifier of the coalition rather than as a divisive wedge" (Campbell, 2006).

Outcomes

How do we know whether environmental justice principles were effective in the highlighted cases? One way we might examine this question is through a close assessment of the outcomes, particularly in light of the previous point of a balance needed between ecological and social goals.

> In terms of outcomes, the Bronx River collaboration has led to the most sweeping changes in land use and the largest dedication of public resources, along with a high degree of public awareness both in the neighborhood and throughout the city.
>
> (Campbell, 2006, p. 34)

The following resulted from the Bronx case:

- Creation of physical space:
  - The Bronx River Alliance (BRA) was able to create 40 new acres of parkland along the river.
  - There was the creation of waterfront and on-water recreational opportunities:
    - For example, BRA create three new canoe/kayak put-ins on the river.
    - It also created a greenway that connects across the city.
- The ecological impact of the restoration has been minimal:
  - Thus far they have engaged in the visual impact piece of cleaning up the river (e.g., trash removal, creating a salt marsh, channel modification).
  - But the river is still polluted.
  - Though there is a full-time team working on this, there is a high cost—over $14 million or more—and it is based on getting volunteers to assist.
  - These projects may need to be scaled up.
- Social impacts can be viewed via the outputs of the group:
  - Large capital funds have been invested into the river.
  - There is a re-granting program that was used by the BRA. Here the BRA would apply for state and federal grants and then would re-grant these to local community-based groups. This was also seen as a successful output.
    - The large coalition of 65 groups was seen as a positive outcome.

While for the other cases outcomes were not as extensive as those seen in the Bronx River case, in both the Chattanooga and Mill Creek cases there

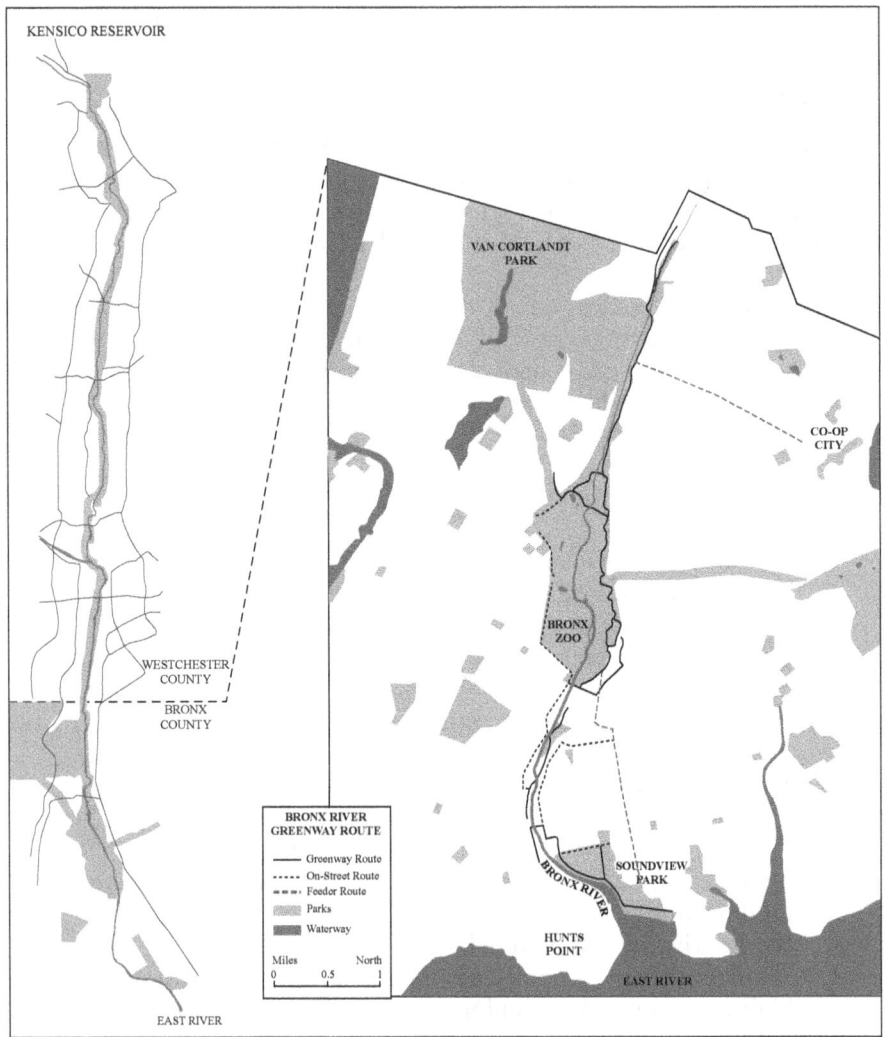

Figure 4.5 Bronx River Greenway
Source: CNY Parks & Recreation, and redrawn by Ryan Mackerer

were outcomes that can be seen as successfully taking environmental justice principles into consideration. In the Chattanooga Creek case, regional and federal actions on pollution have been engaged in the community. As such there has been improvement in neighborhood conditions in health services, educational opportunities, housing, job training and abandoned properties (Rogge et al., 2005). For Mill Creek simply having the creek added back to the watershed maps in 1999 (Spirn, 2005) was considered a major outcome aligning with environmental justice principles as now there was recognition that the creek existed and that it impacted community members' lives, leading to more discussions on how best to address the contamination problem.

"What can be": Intergenerational equity/continuity concerns

Taylor (2002) indicated in her Environmental Justice Paradigm that justice has both an intra- and intergenerational component when dealing with environmental decision-making. Here one can assert that in order for justice to be achieved, participation by those in the current generation must take place, but plans must also be put in place for future generations to be involved in the decision-making process. In doing a comparison of the civil rights movement (CRM) and the environmental justice movement (EJM), Bryant & Hockman (2005) pointed to a lack of students and by extension a younger generation involved in the EJM compared with the CRM. As such there are questions about the posterity and perpetuity of the EJM. It therefore makes sense to ask questions regarding the way in which posterity is seen in river revitalization work as it pertains to who will benefit from the restoration initiatives.

Given that environmental injustices are perpetuated both within and across generations, we postulate that intergenerational coordination might be one way to move toward environmental justice in dealing with river and stream restoration initiatives. In examining intergenerational equity/continuity, two variables are assessed (Table 4.2):

1. Outreach programs
2. Funding implementation for the restoration

Outreach programs: Youth and elder engagement

In looking at intergenerational equity and continuity, the question that must be answered is "What opportunities are available for engaging both younger generations and in some cases elders in the restoration efforts?" Taylor (2002) makes it clear that intergenerational equity is important for environmental justice principles to be incorporated into decision-making. Further, Hillman (2005), in examining perceptions of the Hunter Valley restoration initiatives in Australia, indicated that young persons were concerned about who should be involved in the rehabilitation of the valley.

> The findings of this research challenge stereotypes about young people as uninterested in natural resource management. Acknowledging the views of historically-excluded groups such as young people, while intrinsically just, also provides new perspectives through which a more holistic and balanced view of river use, health and management can be obtained.
>
> (Hillman, 2005, p. 158)

Table 4.2 Intergenerational equity/continuity concerns

Case study site	Funding implementation	Outreach programs
Anacostia River, D.C.	Partnership between different branches of government, NGOs and private citizens	• There is the Anacostia Watershed Citizens Advisory Committee • There are more than ten volunteer groups working with the Anacostia River Restoration Partnership • Groundwork DC focused on youth education
Bronx River, NY	City government	• Education program—targets both youth and adults • Outreach events—for example the Bronx River Golden Ball
Chattanooga Creek, TN	City and federal funds used	
Onondaga Creek, Syracuse, NY	Federal funding used through the Onondaga Lake partnership for developing the revitalization plan	• Extensive citizen outreach, including working group, public forums, surveys, workshops and internet site
Mill Creek, Philadelphia, PA	Grant support from university—community relationship Federal funding provided once the revitalization became a possible option	• Middle school landscape literacy project

There is a perception that decisions about how the resource is used, managed and restored should be made jointly among community organizations, landowners and government. This is one of the main reasons why we need outreach in revitalization projects. Outreach programs have taken multiple forms in the river revitalization projects that have been examined in this chapter.

First, significant education and outreach programs have focused on both youth and general education of surrounding community members. In the Bronx River project, community members could get involved in the river revitalization project in a variety of ways. Community members have engaged in educational programming by involving the main Bronx River Alliance (BRA) group, the parks department and other civil society partners. Examples of these initiatives include the fact that the education team of the BRA runs the Bronx River Stewards program, which focuses on volunteer data collection and monitoring of the river (Campbell, 2006). The BRA has also created a classroom-based curriculum called "Wade in the water" that focuses on water quality, macroinvertebrates, plants, watersheds and environment in the community (see Bronx River Alliance website). Independent nonprofits like Rocking the Boat use the river as a recreation and education site, getting youth out on the water in handmade wooden boats that they create (Campbell, 2006, p. 53).

Similar work is also done for the Anacostia River on two fronts. The first involves a formalized Steering Committee that is part of the leading organization driving the revitalization project (Anacostia Watershed Restoration Partnership, AWRP). This steering committee is seen as the citizen arm of the revitalization project. It provides a vital link between the watershed community and the AWRP to ensure that public interests are considered during all restoration and protection projects and activities (Brendes, 2007; Turner, 2002). The steering committee holds citizens' summits to hear community concerns and engage them in the process (see AWRP website). Second, Groundwork DC partners with other organizations to do specific environmental projects engaging youth on the river. One such project is the Anacostia River Bandalong Litter Trap, a program that removes trash from litter traps along the river. The organization helps maintain the traps along tributaries of the Anacostia. More than ten volunteer groups work on the Anacostia River with multiple opportunities for volunteering. These groups can be seen as a way to engage both young and old in protecting the ecosystem. Further, the Mill Creek landscape literacy project is perhaps one of the most substantive youth engagement projects for river revitalization. As Spirn (2005) describes, this project was a sustained curriculum-based initiative where students from the University of Pennsylvania developed educational materials with middle school students on the history of the creek. Through this initiative students were able to understand the location of the creek and they took an interest in seeking cleanup for the creek.

Another form of outreach that is common among the cases is that of large, one- time annual events that bring the community together. In the Bronx, for example, the

> Bronx River Golden Ball event (Figure 4.6) was created in 1999 and pairs a day of canoe trips down the river with community events and festivals along the riverbanks. Two artists conceived of this yearly event as a way to symbolically unify the river by floating a golden ball down its length. Subsequent others, including The Amazing Bronx River Boat Flotilla, which involves community members and politicians getting out on the river and experiencing it by canoe.
>
> (Campbell, 2006, p. 45)

The Bronx River Bio Blitz has supported the goals of the Bronx River Alliance and strengthened the bonds between the groups working on the river. We do not see these large-scale events across the other river revitalization cases, but such events are good ways of attempting to reach multiple people and multiple jurisdictions. They can promote cross-generational

Figure 4.6 Golden Ball event on the Bronx River
Source: Photo by Anne-Marie Runfola

interactions and help get younger people engaged in the work of revitalization.

Some concerns or words of caution must be highlighted when examining outreach programs as a form of promoting intergenerational continuity. First, using these outreach programs leads us to question whether and to what extent these initiatives are actually engaging those living in and around the river as well as giving a sense of public participation in decision-making. Though in most of the cases there are outreach opportunities for community involvement, there does not seem to be intentional intergenerational collaboration in these projects. Here the claim is being made that there are limited opportunities for grooming the younger generation to take over the leadership of the organizations that engage in the revitalization projects. Finally, while current outreach opportunities can be seen as an attempt to build cross-generational bridges to get younger people involved in the movement, there is limited evidence of elders being actively involved in the outreach programs of these initiatives. In a study, Metzger & Lendvay (2006) highlighted the importance of having elders involved in restoration projects. In the case of the Bayview Hunter Point community, which they reported on, though the community was involved in the restoration project, there was a generational gap in terms of who was actually involved. Most of the participants who were recruited as research assistants (RAs) were younger persons. Some of the elders in the community were vocal about the lack of involvement of elders, indicating that there needs to be a better representation of community demographics in these restoration projects. This is one area where river restoration and revitalization work can improve and take environmental justice into consideration, that is by increasing the number of elders who are directly involved in the planning, implementation and management of revitalization efforts. In order to have intergenerational continuity, efforts need to be more than single-day events. While such events are successful and occur every year they raise questions about the extent to which there is long-term commitment among community members and whether they will feel ownership of the revitalization effort.

Funding implementation

Funding is important for intergenerational equity and continuity. For any social movement to be successful resources must be made available (McCarthy & Zald, 1977), and these resources must be sustained. In these cases there is a combination of funding sources. For example, while the Bronx River Alliance is an NGO, it often works closely with the New York City (NYC) Parks Department, which provides in-kind support for the work of the Alliance. Sources of funding in the Bronx River case are multiple and sustained and include federal sources, foundations and city funds (see Campbell, 2006 for details on funding). Mill Creek, Chattanooga Creek and Onondaga Creek also received federal funding for their respective restoration projects. For Mill Creek the revitalization was wrapped up in the redevelopment of the area.

> The Philadelphia Water Department, Philadelphia Housing Authority and the Philadelphia City Planning Commission submitted a proposal for $34.8 million to the U.S. Department of Housing and Urban Development's Hope VI Program in order to redevelop Mill Creek Public Housing as a demonstration project that would provide an environmental study area for the Sulzberger school and integrate stormwater management measures to reduce combined sewer overflows into the creek.
>
> (Spirn, 2005, p. 407)

For Onondaga Creek federal funding was used directly for revitalization of the creek. The initial cost of the water quality improvements was over $400 million. The county used the federal funding to upgrade the main sewage treatment plant and to build regional treatment facilities that would treat combined sewer overflows during storm events in an attempt to reduce the amount of untreated water going directly into the creek.

With funding and intergenerational equity, though, there are some considerations, as highlighted by the Bronx case. First, the size of the organizations' budgets may not always allow for the revitalization work to be promoted unless they partner with other entities. In the Bronx case, the public–private partnership meant there was a large year-round budget rather than a seasonal one, allowing the revitalization work to continue. In the Anacostia River case, since 1987 there has been formalized cooperation among government agencies that provide funding for the work on the river (Brendes, 2007). Fifteen organizations are listed as part of the partnership on the official website for the Anacostia Watershed Restoration Partnership. It is therefore important for organizations to consider building steady partnerships with government agencies and the NGOs responsible for revitalization work to have a steady stream of funding.

A second funding concern is that access to funds is often limited and competitive among NGOs. This can lead to groups competing for the same resources, which can hamper the revitalization work. One strategy used by the Bronx River Alliance is to encourage collaboration among community groups by seeking out the funding first and then re-granting it to the community groups (Campbell, 2006). The Onondaga Lake partnership after 2000 used a similar process with a mini-grant program from federal and state funds. This allowed only one large organization to compete for the grants while the smaller organizations benefitted by getting access to the resources without the pressure of the competition.

A final concern with funding is considering the role of the private sector in revitalization work. There can be advantages in getting the private sector involved as this frees up public funds for other social services in the community, as was seen in the case of the partnership to construct the RFK stadium along the Anacostia River (Figure 4.7a). However, work with the private sector can be complicated, as private interests do not always match community interests. This was seen in the Anacostia River case where private developers wanted the waterfront to be redeveloped looking at the real estate benefit for the area, but the heart of ecological health and concerns of the river itself were not addressed (Haynes, 2013; Turner 2002) (Figure 4.7b).

In the Bronx case, there was limited to no engagement with the private sector, partly because of the limited local financial resources available within the communities surrounding the most contaminated parts of the river. Additionally for the Bronx River there may have been some deliberate avoidance of working with the private sector given that some of the firms located along the river were waste- and utility-related—the very essence of locally unwanted land uses (LULUs) set up in areas predominantly occupied by people of color. Here are some of the deeper questions organizations should be thinking about with respect to funding and intergenerational equity/continuity:

1. What are the sources of sustained funding to allow the revitalization work to continue across space and time?

Figure 4.7a RFK Stadium and the Anacostia River
Source: Anacostia Community Museum, Smithsonian Institution

Figure 4.7b Anacostia River
Source: Photo by Fritz Reidel

2. How much collaboration should be encouraged across multiple organizations involved in the revitalization work to minimize competition for funding?
3. How much involvement should the private sector have in revitalization projects, and are private and public interests compatible?

Revitalization vs. displacement

One of the recurring themes seen across the cases that has implications for intergenerational equity/continuity is that of the consequence of gentrification as a result of revitalization work. In the Anacostia River case, private investment was encouraged to redevelop the Near Southeast as it was seen as a frontier that needed redevelopment. This community had a rich LGBTQ community nightlife as well as being a predominantly black community. With redevelopment in the area there was some displacement of community members; nightlife for the LGBTQ community was severely impacted and did not recover from this redevelopment work (Haynes, 2013). The redevelopment catered to those living outside the community rather than those within it.

The Bronx River Alliance (BRA) is perhaps a good example of attempting to ensure that displacement does not take place with the revitalization effort. The Alliance has an implementation strategy that addresses economic development and anti-gentrification measures. It put this implementation plan in place by scheduling annual assemblies with all partner groups and ensuring that the organization's board includes members of the community and not only the wealthy. Additionally, one of the BRA's partners has begun work on a project that seeks affordable housing along the river waterfront in an attempt to prevent communities from being pushed out.

The deliberate attempt to avoid gentrification along the Bronx River bodes well for distributive justice, which seeks to allocate environmental goods and bads evenly across groups based on principles of equality and equity (Walker, 2012). In this case the principle of equality is being applied as it is defined, as providing the same thing for everyone. Given that the BRA insists it is making the resource available to everyone, equality seems central. Other revitalization efforts could learn from the Bronx River case how to ensure that community members who experience the environmental harms of a contaminated river are not forced out of the area once the cleanup has started.

Conclusion

Intergenerational equity/continuity is an environmental justice principle that examines the ways in which natural resources and their associated services are used by the current generation and left for use by future generations. River and creek resources have been important for livelihoods, trade and recreation, to name a few of the roles they play. In addition, for many indigenous people, waterways play indispensable roles in their knowledge frameworks, representing values best described as spiritual and cosmological in nature. Historically, the present generation inherited contaminated rivers and creeks as described above for most of the highlighted cases. Despite that, the current generation has found opportunities to engage in river revitalization work as a way to address the contamination and perhaps leave a resource for future generations that allows them to also have access to rivers and creeks. The highlighted cases in this chapter show that environmental justice principles have been incorporated into different aspects of river revitalization. From understanding the neighborhood demographics surrounding the rivers and creeks, to determining what groups are involved in the revitalization process, to the outcomes associated with revitalization, we get a sense of how distributive and procedural justice can be addressed in river restoration initiatives. Examining these cases of river revitalization through an environmental justice lens "permits us to view their struggles through a single analytical lens, as historically and geographically distinct moments within broader processes of colonial domination, racial discrimination, industrial development, environmental transformation, and social struggle" (Perreault et al., 2012, p. 490).

But there are challenges to taking environmental justice into consideration in doing river revitalization work. First, as seen with the Chattanooga Creek case, attempting to work across time is challenging given "corporate and governmental negligence, the lack of local level infrastructure maintenance, the lack of enforcement by environmental regulatory agencies, and institutional discrimination" (Rogge et al., 2005, p. 49). Second, in cases like that of Mill Creek, changes in local management of the project can also hamper progress. There, river revitalization was being pushed through the local middle school near the creek, but once there was new management in the school district, there was disruption to the collaboration between the school and the city to address the concerns of the creek and stormwater management. As a consequence, the revitalization project suffered.

Third, internalized racism can also significantly impact river revitalization projects and therefore intergenerational equity/continuity concerns. The Mill Creek Neighborhood, for example, was called the "Bottom." "Mill Creek is typical of many American inner-city neighborhoods where the residents are predominantly low-income people of color. These 'Black Bottoms' in the U.S. are at the bottom, economically, socially and topographically" (Spirn, 2005, p. 408). Communities often internalize these characterizations of not being a priority for anyone, leading to a lack of efficacy, which diminishes the belief that change can occur within the community. This lack of efficacy can hamper the ability to press for revitalization or continue the revitalization process in cases of disruption in organizations pushing for the revitalization. A final challenge is that intergenerational equity/continuity questions the structure of organizations doing revitalization work and warrants that these structures are stable to allow for intergenerational equity work. For example, in looking at the Bronx River Alliance, the question raised is whether the alliance and its formalization will remain viable for decades to come. Can the alliance structure be replicated in other places (both in the U.S. and globally) as a way to continue to promote revitalization using environmental justice principles?

The cases in this chapter are used to begin the conversation about ways in which environmental justice principles are incorporated into river revitalization efforts. While there are some successful cases where EJ principles have served communities well, there are challenges to ensuring that environmental justice is achieved. What must be noted, though, is that there are benefits to having an EJ focus on river revitalization, the most significant being reducing the impacts of gentrification on urban communities. Given the current need for urban renewal in many spaces and the drive for access to ecological resources in the urban context, river revitalization must continue to make deliberate attempts to incorporate environmental justice principles into the planning, implementation and maintenance phases.

References

American Community Survey (ACS). (2015). Community Facts. American Fact Finder [Website]. Retrieved from https://factfinder.census.gov/faces/nav/jsf/pages/community_facts.xhtml

Brendes, U.S. (2007). Bankside, Washington, D.C. In P.S. Kibel (Ed.). *Rivertown: Rethinking Urban Rivers* (pp. 47–65). Cambridge MA: MIT Press.

Bullard, R.D. (1990). *Dumping in Dixie: Race, Class and Environmental Quality.* Boulder, CO: Westview Press.

Bryant, B. & Hockman, E. (2005). A brief comparison of the civil rights movement and the environmental justice movement. In D.N. Pellow & R.J. Brulle (Eds). *Power, Justice and the Environment: A Critical Appraisal of the Environmental Justice Movement* (pp. 23–36). Cambridge, MA: MIT Press.

Campbell, L.K. (2006). Civil society strategies on urban waterways: Stewardship, contention and coalition building (Unpublished master's thesis). MIT, Cambridge, MA.

Capek, S. (1993). The environmental justice frame: A conceptual discussion and an application. *Social Problems, 40*(1), 5–24.

Davenport, M., Bridges, C., Mangun, J., Carver, A., Williard, K. & Jones, E. (2010). Building local community commitment to wetlands restoration: A case study of the Cache River Wetlands in Southern Illinois, USA. *Environmental Management, 45,* 711–722. doi: 10.1007/s00267-010-9446-x

Ducre, K. A. (2012). *A Place We Call Home: Gender, Race, and Justice in Syracuse.* Syracuse, NY: Syracuse University Press.

Haynes, E.C. (2013). Current of change: An urban and environmental history of the Anacostia River and Near Southeast waterfront in Washington, D.C. (Unpublished senior thesis). Pitzer College, Claremont, CA. Retrieved from http://scholarship.claremont.edu/cgi/viewcontent.cgi?article=1033&context=pitzer_theses

Hillman, M. (2005). Justice in river management: Community perceptions from the Hunter Valley, New South Wales, Australia. *Geographical Research*, *43*(2), 152–161.

Holifield, R. (2001). Defining environmental justice and environmental racism. *Urban Geography*, *22* (1), 78–90.

Honey-Roses, J. (2008). A snapshot of the urban river restoration movement. *Ecology*, *89*(7), 2070–2071.

McCarthy, J. & Zald, M. (1977). Resource mobilization and social movements: A partial theory. *American Journal of Sociology*, *82* (6), 1212–1241. Retrieved from www.jstor.org/stable/2777934

Metzger, E.S. & Lendvay, J.M. (2006). Seeking environmental justice through public participation: A community-based water quality assessment in Bayview Hunters Point. *Environmental Practice*, *8*, 104–114. doi: 10.10170S1466046606060133

Mohai, P., Pellow, D. & Roberts, J.T. (2009). Environmental justice. *Annual Review of Environment and Resources*, *34*, 405–430. doi: 10.1146/annurev-environ-082508-094348

Moran, S. (2007). Stream restoration projects: A critical analysis of urban greening. *Local Environment*, *12*(2), 111–128. doi: 10.1080=13549830601133151

New York City Planning Department. (2015). New York City Population. Retrieved from http://www1.nyc.gov/site/planning/data-maps/nyc-population.page

Park, C. & Allaby, M. (2017). *A Dictionary of Environment and Conservation*. Oxford, UK: Oxford University Press. Retrieved from www.oxfordreference.com/view/10.1093/acref/9780191826320.001.0001/acref-9780191826320-e-4144

Pellow, D.N. & Brulle, R.J. (Eds.). (2005). *Power, Justice and the Environment: A Critical Appraisal of the Environmental Justice Movement*. Cambridge, MA: MIT Press.

Perreault, T., Wraight, S. & Perreault, M. (2012). Environmental injustice in Onondaga Lake waterscape, New York, USA. *Water Alternatives*, *5(2)*, 485–506.

Rogge, M.E., Davis, K., Maddox, D. & Jackson, M. (2005). Leveraging environmental, social, and economic justice at Chattanooga Creek: A case study. *Journal of Community Practice*, *13(3)*, 33–53. doi: 10.1300/J125v13n03_03

Spirn, A.W. (2005). Restoring Mill Creek: Landscape literacy, environmental justice and planning and design. *Landscape Research*, *30*(3), 395–413. doi: 10.1080/01426390500171193

Taylor, D.E. (2002). Race, Class, Gender and American Environmentalism. *USDA Tech Report*. Portland, OR: USDA, Pacific Northwest Research Station.

Trust for Public Land (2002, September). *Tennessee Products Superfund Redevelopment Initiative: Reuse Plans for the Tennessee Products Superfund Site & The Chattanooga Coke State Superfund Site*. Prepared for the City of Chattanooga and the Chattanooga-Hamilton County Air Pollution Control Bureau.

Turner, T. (2002). A tale of two rivers. In T. Turner (Ed.) *Justice on Earth* (pp. 149–163), Earthjustice, White River Junction, VT: Oakland CAS and Chelsea Green Publishers.

U.S. Census Bureau. (2010). Community Facts. U.S. Census [Website]. Retrieved from https://factfinder.census.gov/faces/nav/jsf/pages/index.xhtml

Walker, G. (2012). *Environmental Justice: Concepts, Evidence, and Politics*. New York, NY: Routledge.

Additional useful resources

Anacostia Watershed Restoration Partnership website: www.anacostia.net

Branches, U. (2007). Bankside Washington D.C. In P. Kibel (ed.). *Rivertown: Rethinking Urban Rivers* (pp. 47–66). Cambridge, MA: MIT Press.

Bronx River Alliance website: http://bronxriver.org/?pg=home

Hopkins, A. (2005). *Groundswell: Stories of Saving Places, Finding Community*. San Francisco, CA: Trust for Public Land.

5   Public engagement process

Richard Smardon

Introduction: Challenges to meaningful engagement

Traditionally regulatory agencies have moved forward with structural interventions to address water quality and flood control issues with limited public involvement with local stakeholders (see Chapter 1 for U.S. history and Chapter 2 for European history). This has changed in recent times, but there are still instances where regulatory interests may diverge from local interests (see Platt, 2006; Riley, 2016; Smardon et al., 1996). The other major challenge in public engagement is that the scientific ecologic approach to urban waterways restoration may not appeal to local waterway residents from an aesthetic or recreational access perspective, as Riley (2016) points out with San Francisco Bay examples in her book. The same issue is illustrated in European literature as well (Eden et al., 2000; Eden & Tunstall, 2006). Finally, from an environmental justice (EJ) perspective, we have examples—mostly in the U.S.—of local waterway communities that have little voice in the process (Moran 2007, 2010), have been wronged in the past and have little confidence in local or regional agencies to take them seriously as part of any process. The Onondaga Creek Revitalization Plan (OEI, 2009) is one such example; Near Southside urban creek residents felt that they were victimized by a regional combined sewer overflow (CSO) treatment facility being placed in their neighborhood (Perreault et al., 2012). This made it difficult to move forward with future creek planning because of the residents' distrust of any municipal agency.

Literature review: What has worked and what has not

In her classic 1969 paper, "A Ladder of Citizen Participation," Sherry Arnstein developed a typology of citizen participation that ran from tokenism to stakeholders being involved as co-decision-makers. Jerome Priscolli, a long-time U.S. Corps of Engineers planner, stressed five areas of growing concern in his 2004 paper "What Is Public Participation in Water Resources Management and Why Is It Important?" including:

- Ethical dimensions of water management.
- Water management and civic culture.
- Tension between the technical and political.
- Reconciling the discontinuities between geographic and jurisdictional boundaries; and
- Need for better and more conflict management (Priscolli, 2004, p. 1).

Regional watershed partnerships

We explored the literature on public engagement for water resource management from the watershed scale down to the waterway reach. At the watershed scale a substantial body of literature addresses watershed partnerships (Leach & Pelkey, 2001; Lubbel et al., 2002; McGinnis et al., 1999; Sabatier et al., 2005) with a collaborative approach with multiple stakeholders (Wondolleck & Yaffee, 2000). In Europe the Water Resources Framework has mandated that a collaborative participatory process be used for water resource planning (see Chapter 2 plus Huitema et al., 2010 and Piegay et al., 2002). Four key factors were identified by Leach & Pelkey (2001) for more successful watershed planning efforts from a survey of some 37 published studies: 1) maintenance of a balance between the partnership's resources and its scope of activities, 2) pursuit of a flexible and informal process, 3) various alternative dispute resolution framework variables, and 4) various institutional analysis and development variables (Leach & Pelkey, 2001, p. 383).

In their book *Swimming Upstream,* Sabatier et al. (2005) talk about major themes affecting successful watershed planning partnerships: process and context, participation and representation, civic community, legitimacy and survival. Enserink et al. (2007) stress the importance of understanding cultural factors, while Marks (2004) and Mendez et al. (2012) stress understanding the historical context affecting water resource management. Given that we need more collaborative approaches, Von Korff et al. (2012) pose some basic questions, such as what are the benefits of using participatory approaches and how should these approaches be implemented in complex sociocultural settings to realize them? Such benefits could include better-quality decisions, better acceptance of these decisions and development of social capital. Lamers et al. (2010) assert that a careful process design is needed, and that thorough and continuous stakeholder analysis, use of reflective workshops within and after and use of experienced and qualified process leaders can greatly improve successful outcomes.

There has been little empirical research on what constitutes good or bad participatory process design or how it should be evaluated (Rowe & Frewer, 2000, 2004; Von Korff et al., 2010). From the practitioner perspective one can propose four types of evaluative criteria:

- A good process is credible and legitimate and maintains acceptance of outcomes.

- A good process produces technically competent outcomes.
- A good process has fairness of process; and
- A good process pays attention to educating people and promoting constructive discourse.

Syme & Nancarrow (2002), drawing from Lawrence et al. (1997), propose three forms of procedural justice evaluation measures, which they have applied to the Western Water Authority of Western Australia for treatment and disposal of wastewater. They used simple questionnaire-based indicators for interactive, procedural and distributive justice assessment of the engagement process. According to Syme & Nancarrow (2002:

- Procedural justice "implies that the program is basically unbiased and that an adequate range of participants had a 'voice' or the opportunity to be heard by decision makers".
- Distributive justice implies "adequate representation and satisfaction with the outcome itself"; and
- Interactive justice "implies that the participants have found the process sympathetic to their preferred mode of involvement" (Syme & Nancarrow, 2002 p. 18).

Social learning approaches

In Europe and North America one of the major movements behind collaborative participatory engagement is social learning. According to Mostert et al. (2007), social learning is based on three ideas: 1) All stakeholders should be involved in management and need to collaborate; 2) stakeholders need to enter into a long-term working relationship and organization; and 3) natural resource management is a learning process and requires the development of new knowledge, attitudes, skills and behaviors to address differences constructively, adapt to change and cope with uncertainty (Mostert et al., 2007, p. 19). Mostert et al. (2007) studied some 71 factors that could affect social learning and grouped them into eight areas: stakeholder involvement, politics and institutions, opportunities for interaction, motivation and skills of leaders and facilitators, openness and transparency, representativeness, framing and reframing, and adequate resources. Pahl-Wostl et al. (2007) argue that social learning needs "institutional settings that guarantee some degree of stability and certainty without being rigid." Borowski et al. (2008) note that the European Water Framework Directive participatory structure may not be spatially aligned with decision-making structures, causing a misfit that in turn may impede social learning. This is why Pahl-Wostl et al. (2008) call for the development of cross-sectional capacities and new types of knowledge to respond adequately to the changing dynamics of socio-ecological systems.

New process social learning techniques

As part of social learning, a number of new techniques have utilized co-production. Kuper et al. (2009) utilized co-production of knowledge to make it possible to compensate for the knowledge differential among stakeholders in designing joint irrigation projects in Morocco. Daniell et al. (2010) utilized "co-engineering" for participatory water management processes in Australian and Bulgarian water resource projects. Moellenkamp et al. (2010) utilized a co-design process to establish "niches" as part of the collaborative design and outcomes for the Rhine River basin.

Social learning has also spawned a number of additional participatory approaches, which include:

- Agent based modeling: Whereby the actors' (stakeholders') behavior is represented in the model and so contributes to the modeling process and who later is supposed to use the model(s) for decision-making and strategic planning (Pahl-Wostl & Hare, 2004).
- Citizen juries: Citizen juries were used to increase cognitive learning for water resources management for the Dutch part of the Rhine river basin. There was a high level of cognitive, normative and relational learning levels for jurors but not so much for the policy-makers (Huitema et al., 2010).
- Art-based engagement: Creative writing workshops in a former industrial area in northern England complemented collaborative planning to reimagine a river (Selman et al., 2010). (See Chapter 9 of this volume for more examples.)
- Interactive dynamic modeling: This could be used to facilitate exploration, analysis and synthesis of alternative designs, plans and policies (Bots et al., 2011; Costanza & Ruth, 1998; Loucks et al., 1985). Care has to be taken to involve stakeholders with the modeling process from beginning to end or the model may not be accepted or trusted (Manno et al., 2008).
- Bayesian networks: Bayesian networks were used to structure discussions and visual connections as part of the Upper Guadiana Basin Water resource management planning in central Spain (Zorrilla et al., 2010).

River corridor scale and waterway stretch engagement

As we move to the urban river corridor there is very little literature on participatory engagement of stakeholders. Much of the U.S. literature on participatory process is focused on wild and scenic river development and management (USDI-NPS, 1988). McGowan et al. (2015) advocate utilizing societal objectives to achieve ecological improvements plus the use of adaptive management, education and outreach to maintain long-term engagement. Eden & Tunstall (2006) warn about the "deficit model," where public understanding of what is possible and lack of fit among restoration expectations, policy planners and local publics may pose engagement impediments. In the case of urban waterway revitalization and enhancement, Holt et al. (2012) argue that network analysis is needed for multi-stakeholder collaboration for sustainable management of urban river corridors, especially for developing a shared vision for what is needed. Hagerman (2007) reinforces this need for urban networks and suggests that new governance regimes are needed for visioning, planning and development of such urban waterway neighborhoods and districts.

Specific urban waterway engagement processes include:

- Developing landscape literacy: With her work in Mill Creek, Philadelphia, Anne Spirn (2005) has used a process of developing local landscape literacy as a means of recognizing and redressing past injustices through urban planning and designs for future community development. See more details in Chapter 4.
- Future visioning: Selman et al. (2010) have utilized a series of creative writing workshops to foster changes in knowledge, attitudes and actions about the use and management of urban river environments in a former industrial area in northern England.
- Building watershed narratives: Yocom (2014) has utilized a narrative approach as a chronotype for untangling the biophysical and sociocultural complexities of an urban stream restoration process in Seattle, Washington.
- Place-based conservation: Wessells (2010) argues for building capacity for equity in urban watershed management via experimental, symbolic and identity-based means. Her work draws from fieldwork in Los Angeles, California.

By far the most developed process for civic engagement has been done by Judith Petts (2006) as it is applied with the SMURF project in the Upper Tame catchment in West Midlands, U.K., as part of the urban area within the City of Birmingham. She describes an engagement process that makes technical knowledge publicly accessible and also translates practical questions and public problems into an expert discourse. Her detailed engagement process is described in the following case study.

Urban waterway case studies

Upper Tame Catchment in Birmingham, U.K.

The Sustainable Development of Urban Rivers and Floodplains Project (SMURF) was funded by the European Union (EU) Life Program. As part of this project there was a specific river restoration project within the city of Birmingham, U.K. (Figure 5.1). A learning approach was utilized for

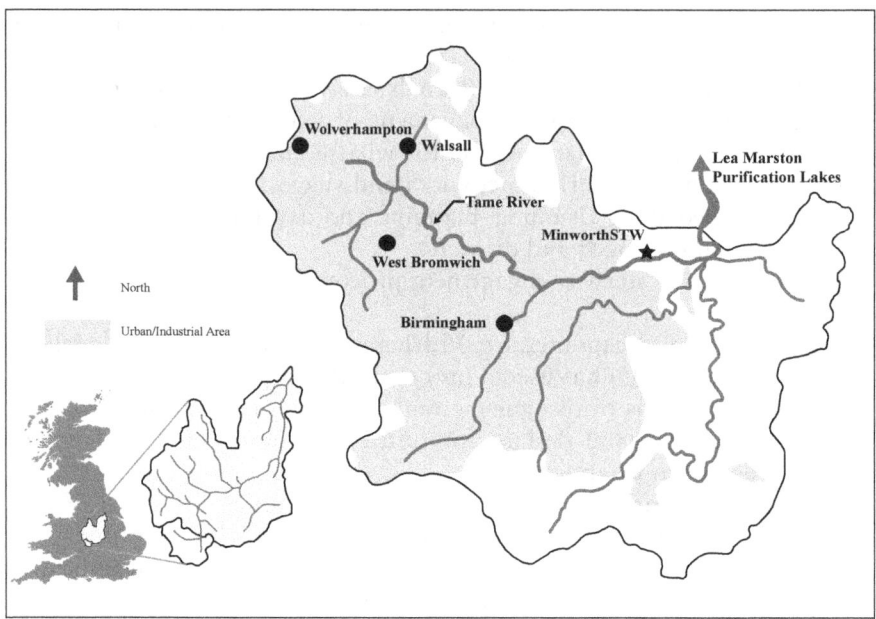

Figure 5.1 Upper Tame catchment
Source: EA (2013), redrawn by Ryan Mackerer

engagement, which emphasized decision-making that was place-based, cooperative and inclusive of multiple parties. The river corridor scope of work was enhancement or rehabilitation focused rather than restoration focused. A geographically whole city perspective was used to create a vision. A second or small-scale project focused on the Perry Hall Playing Fields adjacent to the river.

The primary form of engagement was discussion workshops where a small number of participants (20 per group) engaged with experts and decision-makers. These participants were recruited to represent a broad range of interests and managed to optimize engagement of people who were informed but not experts. A community assessment process was utilized via key organizations, internet and local press and library searches to generate names for participants for the citywide visioning as well as the Perry Hall Playing Fields project. Petts (2006, 2007) notes that young people under 18 and immigrants were difficult to recruit.

A number of specific engagement techniques were utilized, as one can see from Table 5.1. These include:

Table 5.1 Requirements of key engagement management elements to support learning

Management element	Requirements
Recruitment of representative interests	Locally informed perspectives from a variety of viewpoints
	Sufficient time for recruitment
	Recruitment by lead facilitator to provide contact continuity
	Direct contact with potential participants
	Core participants engaged through the whole process
	Information provision to wider community
Active facilitation	Independence from project decision/delivery agencies
	Act in interests of lay and expert participants
	Control the more dominant voice while encouraging the weaker
	Significant facilitation experience
	Subject knowledge and ability to synthesize technical information
	Maintenance of balance between assistance and direction
	Assist discussion by elucidating issues and making essential linkages
	Ongoing participant contact within and outside of meetings
Collaborative framing	Achieve buy-in by showing issue framing is not closed down
	Agreed-upon terms of reference and ground rules
	Time to explore all issues but ensure focus on what is possible
	Continuous use of narrative and visual prompts
	Capitalize on and be seen to value local and experiential knowledge
	Co-produced lay and expert framing and priorities for action
	Mechanism to ensure that official agencies recognize all local issues
Optimizing interaction	Project team pre meetings
	Making technical presentations publicly understandable
	Bringing public concerns into an expert discourse
	Expert and public informal and formal interaction throughout
	Continuous individual expert involvement
	Site visits
	Background information provision
	Small-group and plenary discussions
Managing the unexpected	Sufficient funds to allow flexibility of process
	Facilitator close monitoring of process
	Open communication when problems arise
	Manage expectations to maintain confidence and build trust

Source: Petts (2007, p. 178)

- Active facilitation with two independent facilitators with careful balance between assistance and direction.
- Collaborative framing so nothing was predetermined for the vision city-wide or the specific project.

- Optimizing interaction by balancing public concerns with the expert discourse while providing technical translation for lay participants.
- Social learning through provision of information and management of discussions with a mix of plenary and small-group discussions.
- Developing a shared vision even with differing priorities and reaching consensus by agreeing on a balance of characteristics; and
- Managing the unexpected through adaptive process management.

Key lessons learned from the SMURF project in Birmingham include:

- "Learning as a balancing act requiring careful management of the powerful framing effects that can privilege expert knowledge while ensuring that local knowledge with public issues and priorities are balanced against what is practically feasible" (Petts, 2006, p. 178); and
- The two-stage process was very important as it allowed quality time for expert-lay interaction and enabled co-construction of the problem and definition of community priorities and technical principles.

Onondaga Creek Conceptual Revitalization Plan, Central New York, U.S.

This case study describes the development process for the Onondaga Creek Conceptual Revitalization Plan (OCCRP). For the Onondaga Creek watershed and creek corridor the major environmental justice issues have been addressed by Perreault et al. (2012) but will be summarized here. For context, the watershed in the upper reaches includes rural and suburban communities; the Onondaga Nation lies in the central part of the watershed; and the lowest part of the watershed includes urban neighborhoods in Syracuse's Valley and Southside districts (see Figure 5.2). There are two primary communities with EJ issues: the Onondaga Nation and Southside neighborhood along Onondaga Creek (see Figure 5.2). Historically the Onondaga Nation depended upon abundant aquatic, riparian and upland environments associated with Onondaga Lake and Creek prior to Euro-American settlement, which began in the late 18th century. Colonization transformed the watershed irrevocably over 200 years, through salt mining, Euro-American dam construction (including upon sovereign territory of the Onondaga Nation) and profound industrial and municipal pollution. Further, there are unique related impacts seen nowhere else, such as the mudboils, which have deposited tons of sediment into Onondaga Creek upstream of the Onondaga Nation, impairing aquatic habitat and transforming (and harming) the relationship the Onondaga Nation's people have with the creek (Perreault et al., 2012). The Near Southside community had been subjected to forced housing within a high-risk location in the Onondaga Creek floodplain and subject to combined sewer overflow events along the creek. The OCCRP effort had to overcome both communities' historic environmental justice issues during the planning and engagement process.

Figure 5.3 illustrates the components that make up the OCCRP project.

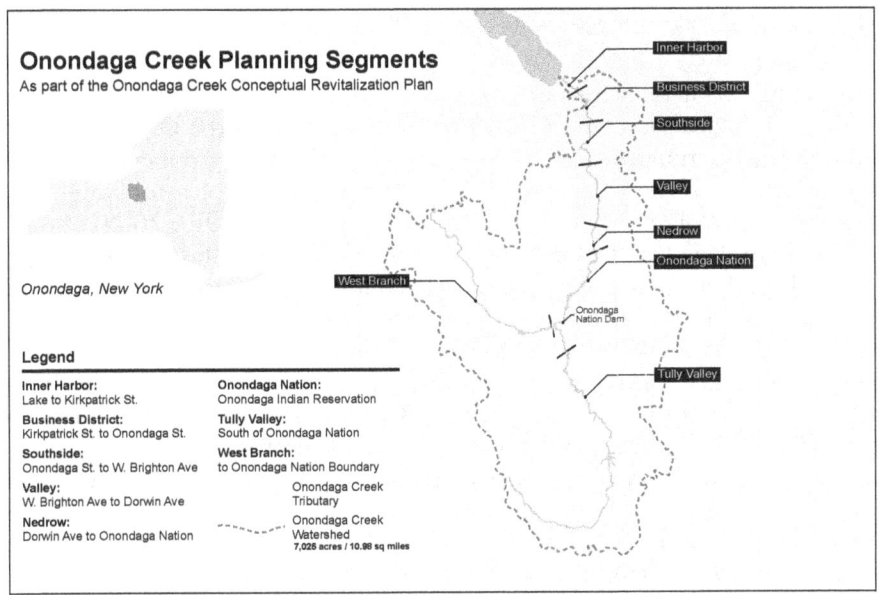

Figure 5.2 Onondaga Creek watershed in central NY USA
Source: Mark Warfel Jr.

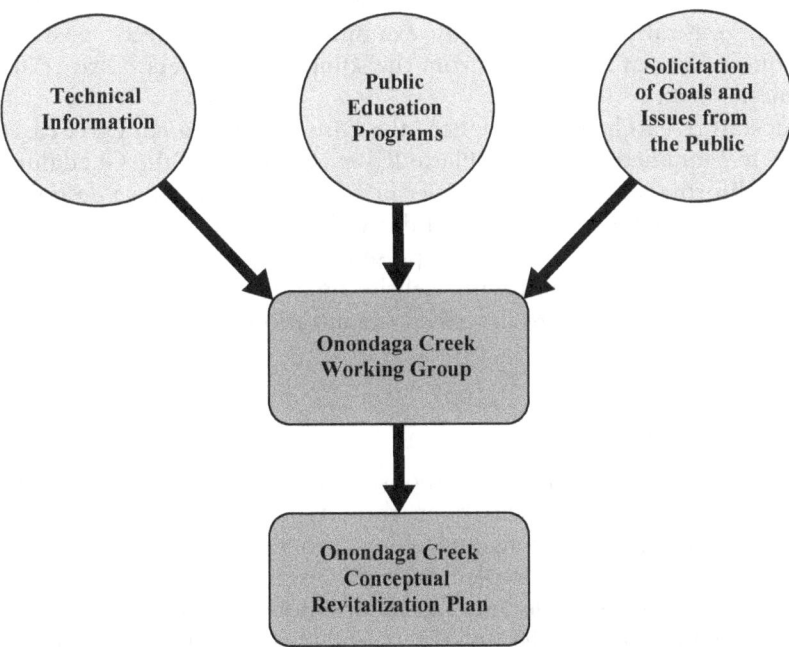

Figure 5.3 Onondaga Creek Conceptual Revitalization Plan project components
Source: Permission obtained from OEI (2009) and figure redrawn by Ryan Mackerer

Technical information process

A compilation of relevant background information concerning the watershed was a logical first step toward the development of the OCCRP. Three reports were completed: a summary document describing the current state of Onondaga Creek, a description of case studies of successful watershed restoration and planning (OEI, 2008), and the OCCRP (OEI, 2009).

The Onondaga Creek Fact Sheets describe the current state of Onondaga Creek. To produce the fact sheets, staff conducted literature searches and compiled relevant information into documents based on topic areas. Project staff initially developed a broad list of topic areas. These were then reduced based on material found in the literature search and Dr. Richard Smardon's judgment of what the Onondaga Creek Working Group needed to know to develop the plan options, in his role as group facilitator.

Once prepared in draft form, the fact sheets were used as an interactive planning tool with the Working Group, which reviewed and critiqued each sheet. Revisions were incorporated into the Fact Sheets and revised. The Fact Sheets were used by members of the Working Group to deepen their understanding of existing conditions (reinforcing that learning via field trips and guest speakers) and to develop options for the revitalization plan.

The *Case Studies Guide: Conceptual Alternatives to Onondaga Creek* (OEI, 2008) was developed to provide the community and decision-makers' examples of stream revitalization throughout the country. Each river is unique; no single example will provide a perfect reference with which to guide local restoration. However, by examining many projects, answers to local questions can be gleaned from the solutions of others (Otto et al., 2004; Riley, 1998).

Project staff researched and produced the *Case Studies Guide*. Three cases were closely examined: the South Platte River in Colorado, the Guadalupe River in California and the Bronx River in New York. Each case describes river history and current projects, and draws lessons for Onondaga Creek revitalization. Twelve short cases are presented, emphasizing one or two salient revitalization examples with website links for further exploration. Some of these case studies are described organizationally in a later part of this chapter.

Public education process

Two project team members were responsible for conducting public education programs, and they focused on three types: stewardship-building events, educational presentations and school programs. A local parks and green space advocacy group complemented the programs with its own event in 2005. All programs were designed to occur before and during the public forum phase and foster public awareness and involvement in Onondaga Creek watershed issues. Adult-oriented programs were also intended to build awareness of and encourage involvement in the plan development process.

Goal and issue solicitation process

The project team devised two methods to gather goals and concerns: community forums and stakeholder organization meetings. The participation goal was to assess the larger watershed community's visions and concerns for Onondaga Creek, which in turn would assist the Working Group in their development of the revitalization plan. Gathering public input prior to the development of the plan allowed themes and goals important to the community to be incorporated into the plan (Firehock et al., 2002). Figure 5.4 was used at the community forums and stakeholder organization meetings to explain what would happen to the input of meeting participants.

There were several rationales for gathering public input prior to plan development. First, developing the OCCRP was to be a lengthy process. Few citizens would be able or willing to fully participate in years of meetings for plan development. However, many more people could be reached in one-time meetings in formats designed for larger groups. These meetings served the purpose of developing visions and priorities (Innes & Booher, 2000). Second, implementation of the OCCRP is voluntary. Voluntary plans need support and involvement of stakeholders throughout the process, both to develop a sense of ownership and to increase the chance of implementation (Scholz et al., 2002; Smolko et al., 2002).

The Onondaga Creek Community Forums were designed to draw goals and issues from watershed residents and other interested individuals. The meetings were open to the public and marketed as such, through community outreach efforts including public service announcements; newspaper stories; flyer distribution in targeted neighborhoods via community groups and libraries; "get the word out" kits distributed via email to local organizations (this consisted of a flyer, project information documents and suggested text for newsletters and email notification); community calendars available in the newspaper, television and online; press releases; and media kits to the local press. The U.S. EPA's *Getting in Step: A Guide to Watershed Outreach Campaigns* (USEPA, 2003) inspired many of these methods of communication.

Forum locations were distributed within the watershed geographically and according to population density. Location choice was constrained by size, configuration, parking, availability of facilities perceived as accessible and recognizable to the community, and by the need for facility fees to be within the project budget. Five forums were held in the City of Syracuse; two were outside the city. Three types of input were collected from participants at the forums: dot board results, verbal comments (scribed to flip charts) and written responses (from question cards). Dot board data were entered into Microsoft Excel. OEI staff entered verbatim input collected from the flip charts and question cards into a Microsoft Access database. Verbal and written inputs were based on the open-ended questions.

Topics most frequently mentioned in aggregate for the community forums were obtained from written cards completed by participants at each meeting.

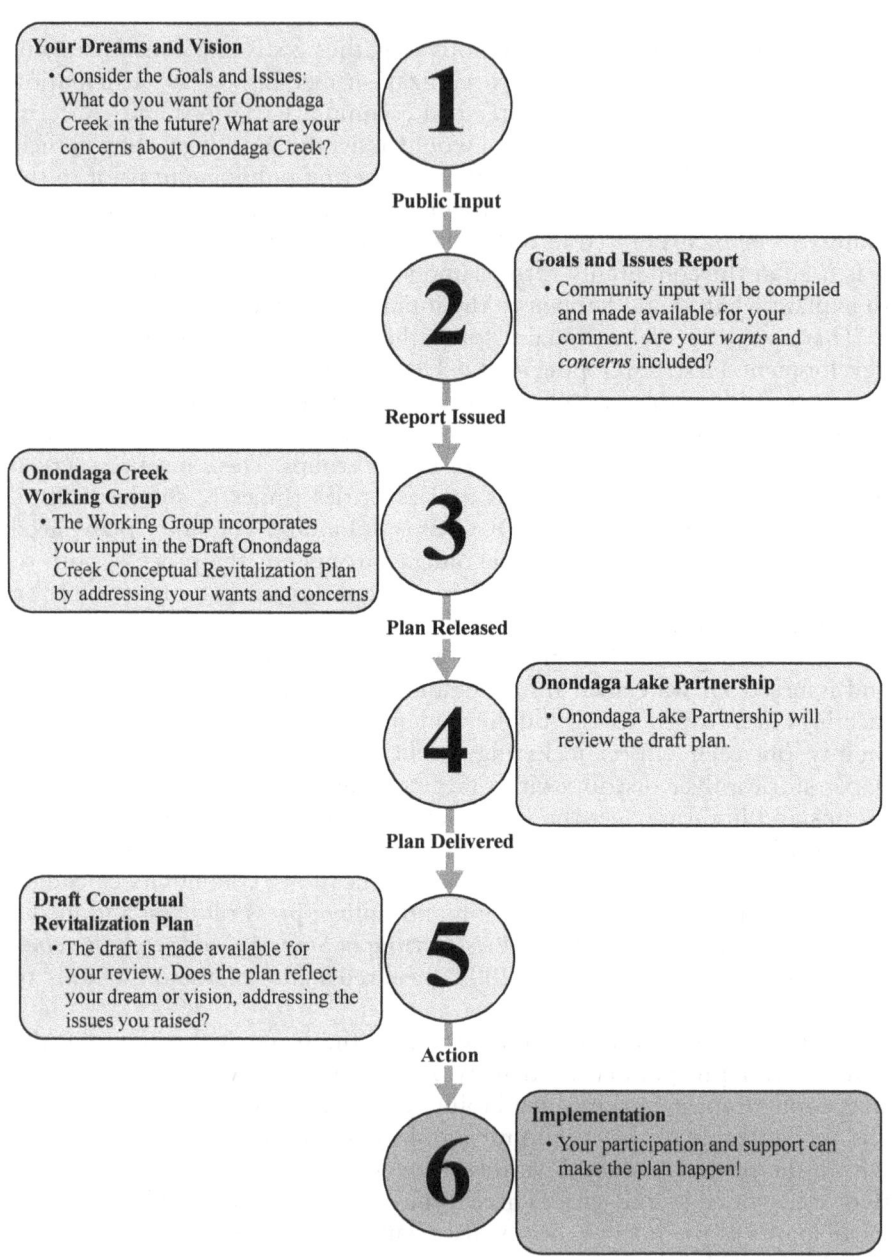

Your Dreams and Vision
- Consider the Goals and Issues: What do you want for Onondaga Creek in the future? What are your concerns about Onondaga Creek?

1

Public Input

2

Goals and Issues Report
- Community input will be compiled and made available for your comment. Are your *wants* and *concerns* included?

Report Issued

Onondaga Creek Working Group
- The Working Group incorporates your input in the Draft Onondaga Creek Conceptual Revitalization Plan by addressing your wants and concerns

3

Plan Released

4

Onondaga Lake Partnership
- Onondaga Lake Partnership will review the draft plan.

Plan Delivered

Draft Conceptual Revitalization Plan
- The draft is made available for your review. Does the plan reflect your dream or vision, addressing the issues you raised?

5

Action

6

Implementation
- Your participation and support can make the plan happen!

Figure 5.4 What happens to my input?
Source: With permission of OEI and figure redrawn by Ryan Mackerer

All written input, catalogued according to goals or concerns, was analyzed and assigned a one- or two-word code, identified as a key word that captured the contextual meaning. Key words were generated based on review of the data rather than created beforehand. The input was grouped by key word for each forum and sorted by frequency. Frequencies were aggregated across forums. Input was then graphed by most frequently occurring key word. This process was influenced by two methodologies for analyzing qualitative data: content analysis (see for example USEPA, 2002) and grounded theory (see for example Silverman, 2003 and Strauss, 1987).

The second type of meeting, the stakeholder organization meetings, were intended to draw goals and issues from members of organizations, institutions and businesses—in other words, groups that would have a particular interest in Onondaga Creek revitalization. To determine meeting format and groups to approach, OEI staff gathered advice from several community leaders in government, nonprofit and business roles.

Eight stakeholder organization meetings were held; the majority occurred in the first half of 2007. Six small meetings were distributed among civic and environmental groups with existing meeting schedules and two large meetings were conducted. About 120 individuals representing more than 60 organizations attended the two large stakeholder meetings.

Project staff communicated to the Working Group the top themes from the community forums and the stakeholder organization meetings in fact sheet format. The Working Group also received copies of community forum written input and assisted in categorizing data into themes. Most Working Group members gained firsthand experience with the community's goals and concerns by attending both types of meetings. Subsequently, the Working Group and project team incorporated community input into the plan development process, as described in the next section. Figure 5.5 illustrates the goals and issues solicitation process.

Working Group process

The Onondaga Creek Working Group met monthly from February 2005 to September 2008. Working Group participants were recruited to represent a variety of interests and geographic areas of the Onondaga Creek watershed. Meetings were held monthly, on the first Wednesday evening of the month. All meetings were open to the public. Informal methods of notification about Working Group meetings were used on occasion, particularly handouts and posters distributed at local environmental events and meetings.

Learning phase and plan components development

As preparation for developing the revitalization plan components, the Working Group engaged in a learning process about the Onondaga Creek watershed; members informed one another as they shared information and experience. Additionally, the Working Group added to their existing

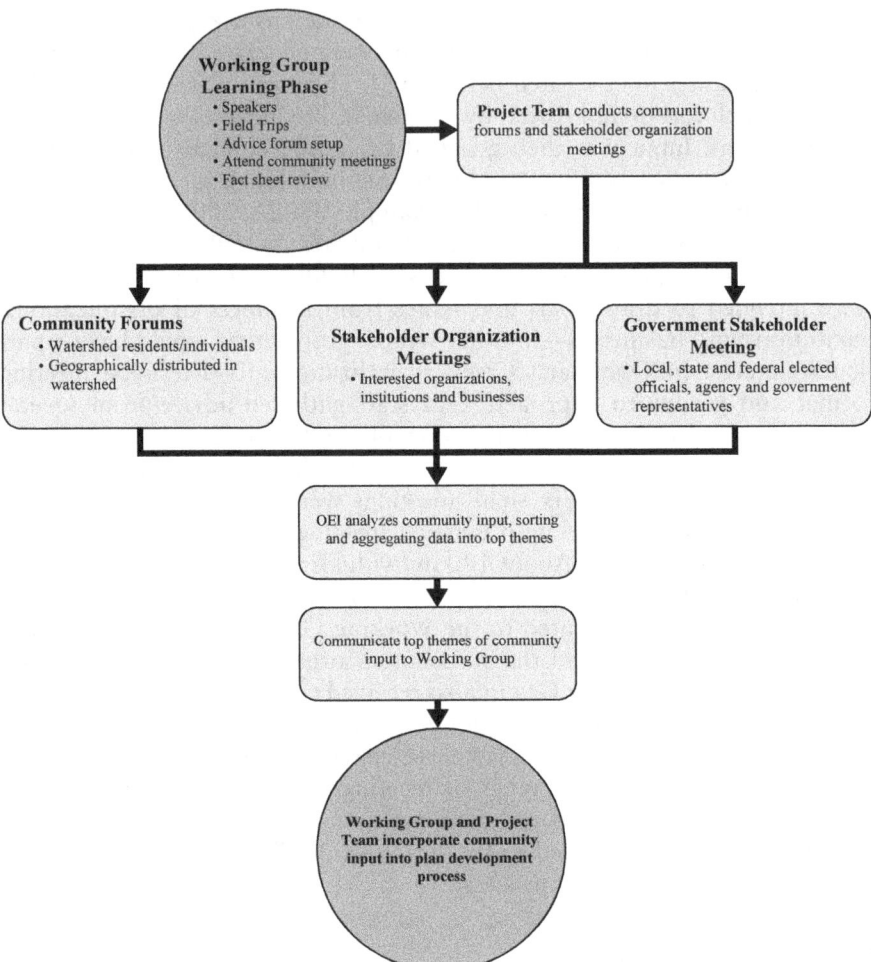

Figure 5.5 Onondaga Creek Revitalization Plan outreach strategy
Source: With permission of OEI (2009), redrawn by Mark Warfel Jr.

knowledge by learning from guest speakers at Working Group meetings, selecting and participating in creek-themed field trips, participating in the goals and issues solicitation process and reviewing the Onondaga Creek Fact Sheets.

After the fact sheet review, the Working Group developed the components of the OCCRP. First, the Working Group developed and refined drivers, the driving forces or motivators, for revitalization. Next, revitalization options for Onondaga Creek were developed through a series of meetings devoted to specific topics: hydrology, biology and land use/access/recreation. The project team invited local scientists and practitioners as resource experts in each topic area to advise the Working Group during options development. The resource experts included individuals from local universities and government agencies. With options complete, the Working Group completed a design charrette, a planning exercise where ideas for revitalization were placed on a series of maps over two intense sessions. This process is described further in Chapter 7.

Other urban waterway organizational case studies

The following is a series of urban waterway restoration/revitalization studies that illustrate different decision-making and organizational models. There are some common themes throughout as well as some unique organizational schemes.

Bronx River restoration case study

The magnitude of impact here is larger than that of Onondaga Creek because of the dense urban environment and greater population in the Bronx. The overall length of the Bronx River is 23 miles; about 3 miles shorter than Onondaga Creek, but there are more sources of pollution in New York City. A key source that documents the Bronx River history is Hopkins' (2005) book *Groundswell* and Jill Weis, who produced Table 5.2.

Different impetus: Historic disregard for the water body by public officials is a factor common to both the Bronx River and Onondaga Creek. After 30 years Bronx environmental and community activists have transformed a grassroots effort into a major river stewardship organization that eventually garnered official support from New York City. The attention of community activists and local citizen groups was drawn to Onondaga Creek by several Onondaga County CSO control projects that were negotiated via federal court as a result of a citizens' lawsuit over Clean Water Act violations in Onondaga Lake.

Ad hoc working groups as incubators for more formal organizations: The important transition point for the Bronx River project was the formation of the Bronx River Working Group, which created the Greenway and Restoration plans, and then re-formed as the Bronx River Alliance to implement projects and advocate for the river. Overcoming the Working Group inertia and using it to establish a more formal organization is a lesson to be learned from the Bronx River case study.

Access to the water: A lesson from the Bronx River case study, and from most others, is to gain access to the water. A barrier to public access, whether a fence or a highway, isolates people and prevents recreational use. A sense of ownership is lost when there are such obstacles, whereas contact with natural waterways tends to foster different attitudes. Some of the first

Table 5.2 Bronx River corridor historiograph

Bronx River hits rock bottom	Early 1970s	Lowest point—active trash and car dumping, high crime in vicinity of river, vast algal blooms
Bronx River Restoration Project founded; symbolic action and ownership events	1974	Bronx Borough Police Commander Anthony Bouza (note spelling) organized community meetings to address the river's social and ecological health. Ruth Anderberg organized cleanups in the West Farms area of the river and founded the Bronx River Restoration Project (BRRP)
	1980	Bronx River Master Plan created
	Early 1980s	Symbolic action events such as park adoption and street festival/demonstrations like the "Not In My Neighborhood You Don't" events led by neighborhood matron Juanita "Ma" Carter inspired even more interest and ownership of the area. Ma would later be awarded a Park Warden title for River Park. Anderberg formed the Youth Conservation Corps (not to be confused with an NPS program of the same name) that spearheaded the Bronx community garden movement that persists today. In-kind donations for cleanup efforts become more common; an example is when the NYC Parks and the National Guard donated personnel and equipment to remove abandoned cars from the river
CSOs become priority; cross agency involvement	1986	Nancy Wallace took over leadership of BRRP, forging a relationship with NYC Parks Commissioner Henry Stern and Mayor Ed Koch to commit funds for ecological restoration, resulting in some small grant-supported successes
	By 1990	Combined Sewer Overflows or CSOs identified as the river's most serious contemporary threat. Pressure applied to NYC Department of Environmental Protection by BRRP, other partner organizations and the Bronx Council for Environmental Quality

Table 5.2 (Cont.)

Bronx River hits rock bottom	*Early 1970s*	*Lowest point—active trash and car dumping, high crime in vicinity of river, vast algal blooms*
Bronx River Working Group founded— drastic increase in river programming	1997	Jenny Hoffner convened 60 organizations and formed the Bronx River Working Group. A cohesive Bronx River Greenway Plan and Bronx River Restoration Plan were created that consider the corridor as an entire system. Funding support from the USDA Urban Resources Partnership Program, the U.S. Department of Transportation and the EPA Wetlands Protection Program supported this effort. Some key organizational partners include: • The POINT • Sustainable South Bronx • Rocking the Boat • Youth Ministries for Peace and Justice • U.S. Army Corps of Engineers • NYS Department of Conservation • Westchester County Department of Environmental Planning • New York City Restoration Project • Fordham University Community Service Project • + others
	1999	Current vision of the Bronx River Greenway was formed
	2000	First Amazing Bronx River Flotilla
The Bronx River Alliance	2001	Bronx River Alliance (BxRA) formed out of BRRP; Hoffner steps down from leadership (characteristics of BxRA available)
	2016	1,000 volunteers donated time, including 6,466 program participants, 741 bags of trash removed, 2,746 trees planted Bronx River Watershed Summit: Dozens of organizations from the watershed met to discuss the state of the river and plan actions. (A list of presenters is available.) New BxRA headquarters under construction; the River House is due for completion in 2018

Source: Jill Weis

people to promote cleaning the Bronx River in the 1970s sought recreational opportunities such as fishing, boating and bird watching. As contact with the river increased, so too did public pressure for improved water quality. Over time, more and more people took enjoyment, ownership and eventually pride in the river. People began to protest the river's condition and petitioned for cleanup and restoration only after the river was used as a recreational resource. But using the river as a living classroom also provided impetus for cleanup (Runfola & Weiss, 2007). Thus the spiral of decline was reversed.

Summary and lessons learned

Government support is important: Public officials and citizens must work together toward common goals. Some politicians will aid a cause because they are interested in the issue; all politicians will respond to public will and political force. Progress on the Bronx River was slow until the city administration realized in the mid-1990s that the public saw restoration as an important and enduring issue.

Continuity of effort is essential: For the Bronx River, different organizations sequentially directed river cleanup and restoration efforts, so there was always a passing of the organizational reins during the transition. This important member continuity sustained consistent pressure to improve the river. Institutional knowledge and influence were not lost during reorganization.

River renewal and urban renewal go hand in hand: The Bronx River runs through some of the most economically depressed neighborhoods in New York City. With little disposable income for travel, local residents see the river as a critical natural asset, providing it is accessible and safe. Efforts around the Bronx River have affected nearby communities. Neighbors along the river have recently repaired abandoned buildings to improve residential neighborhoods. Property values are increasing, crime is falling and people and investors are moving back to the Bronx. Economic and social improvements are potential secondary and tertiary benefits of river or creek restoration. The repayment cannot be estimated in dollars alone. This is one of the major examples of addressing both procedural and distributive EJ issues as covered in Chapter 4.

Wildcat-San Pablo Creek case study

The Wildcat-San Pablo Creeks in North Richmond and San Pablo, California, were perhaps one of the earliest urban stream restoration projects in the U.S. and certainly in California. A good overall account of this case study can be found in Riley (1998). According to Alan LaPointe (personal communication, May 2008), a local activist, it all started in 1980 when the Friends of Wildcat Canyon was formed to protect Wildcat Canyon open space. This group worked with county parks to help acquire lands for parks in the upper Wildcat Creek area. In January 1982 there was $2 million in flood damages in San Pablo, mostly silt damage and downed trees from 1,100 to 1,200 cu ft./sec flows. LaPointe had previously made contact with the famous hydrologist Dr. Luna Leopold at the University of California at Berkeley, and through the Friends of Wildcat Canyon's (FWC) previous work with the county agency. The FWC sought to work out a compromise on the flood issue. LaPointe joined forces with Ann Riley (then a graduate student working with Leopold) to do basic measurements of the creek area and organize local advocacy groups, including some powerful environmental advocates in the Bay area.

Negotiated planning process

A series of meetings were held with major stakeholders over two to three years with facilitation from the county planning and development department. The county water district already had a plan that was heavily engineered to alleviate flooding via channelization. The advocacy coalition came up with a new alternative plan with funding pledged by area environmental groups and the county park district. The alternative plan was much less structure-oriented and more focused on restoration. Eventually the alternative plan was funded through a California legislative initiative.

Ann Riley (personal communication, 2008) feels this was really an environmental justice case as the residents most affected were predominantly minority and lived in very poor neighborhoods, and it was difficult at that time to obtain federal funding as a match of local funding was dependent on the relative wealth of local resources such as property values (Riley, 1989; A. Riley, personal communication, May 2008). Also, according to Riley, there was need of a phase II project for Wildcat Creek because of low flows causing siltation in the streambed. The objective was to configure the channel to reduce siltation buildup.

Lessons learned

With strong advocacy, even disadvantaged neighborhoods can negotiate alternative creek restoration plans. Follow-through to implementation is dependent on strong continuous legislative connections. Adaptive management is needed for low-flow streams that cannot effectively move material through the system.

Guadalupe River, San Jose, case study

According to the former director of the Friends of Guadalupe (personal communication, May 2008), there was at least a 12-year history to the Guadalupe River restoration in San Jose, California. Since the 1940s there had been a number of recurrent floods and flooding studies by the U.S. Army Corps of Engineers (USACOE). In the 1990s there was authorization and funding from USACOE, the Santa Clara Water District (SCWD) and the City and Development Agency of San Jose for the first two segments of a multipurpose project for flood control, habitat restoration/enhancement, recreation and park creation.

Negotiated alternative development

Kathy Muller was the first employee of the Friends of Guadalupe, a non-profit organization formed from several environmental groups in the area. They threatened a lawsuit in 1992 to prevent segment 3 of the Guadalupe River Project under the Endangered Species Act due to risk to steelhead trout habitat. Muller and two prominent board members performed shuttle diplomacy with the City of San Jose, USACOE, SCWD and the Attorney for the Natural Heritage Institute (lead protagonist for steelhead trout). As a result, SCWD, the City of San Jose and USACOE hired an environmental facilitator from CONCUR plus staff from the three organizations to support meetings every two weeks for two years. The Guadalupe Task Force included representatives from neighborhood communities, user groups and companies with vested interests. Over the years of negotiated meetings the task force kept links to congressional representatives to ensure funding if and when an agreement was reached.

A collaborative approach was developed with research agencies, including the U.S. Fish and Wildlife Service, National Oceanic and Atmospheric Administration (NOAA) and National Highway Institute (NHI) as well as SCWD, USACOE and City Redevelopment of San Jose. In one year, the design was radically changed to stop removal of vegetation to keep temperatures low for steelhead trout habitat. A covered bypass design was used to handle water volume, flood control, 1 percent low flow plus opportunities for parks in downtown San Jose. By 2005 the project was built.

According to Muller (personal communication, May 2008) the change in environmental laws plus a project focus beyond only flooding forced USACOE and SCWD to look beyond flood protection, and the health of the fish population became the major driver. The Friends of Guadalupe used newsletters and community meetings as well as the master plan development process to reach out to the community. The Friends of Guadalupe became Friends of Guadalupe River Park and Gardens, a public–private partnership to develop parks, paths and development opportunities along the river within the city.

Moving toward adaptive management

As a condition of the phase 3 planning process, a detailed mitigation/monitoring plan was negotiated that included specific indicators for riparian species, aquatic ecology, fisheries, substrate, etc. The major parties also committed to adaptive management with yearly resultant recommendations.

According to Al Gurevich (personal communication, May 2008), unit manager for the Guadalupe River Watershed, in 2005–2006 a new paradigm of "geomorphic correctness" was proposed by the resource agencies and environmental groups. SCWD was frustrated because a new geomorphic regime, if installed, would cause conflict with existing water structures such as the bypass in place, and might lead to loss of riparian vegetation. SCWD is continuing to collect data but is concerned about shifting design/ management approaches engendered by adaptive management.

Lessons learned

Key players were the Friends of Guadalupe (FOG), the Santa Clara Water District and the City and Redevelopment Agency of San Jose. This is another example of a negotiated planning process initiated by an advocacy organization (FOG) in reaction to a structural alternative. Clearly, this has been a major success from an economic revitalization perspective, bringing major firms back downtown, but maybe not so much from a social perspective of serving disadvantaged communities. This is not a strong EJ case except for the fact that it provided river-related green space throughout the city and economic revitalization of downtown San Jose.

South Platte River in Denver, Colorado, case study

The South Platte River in Denver is a legendary three-part story of river revitalization within an urban context. The first part was captured by Joe Shoemaker (1981) in his book *Returning the Platte to the People*. This book reviews the detailed history of the formation of the Platte River Development Committee at the prompting of Mayor William McNichols in 1974. Until about 1982 this development committee, with little direct power and with representation from Denver neighborhoods and the business community, moved forward quickly. The committee planned projects in river segments, obtained funding to acquire land and building projects and in general believed that having something physical for people to see and experience would fuel demand for more of the same. Substantial progress was made for several parks and trail systems to provide Denver residents physical and aesthetic access to the river during this period.

The year of the river

A second round of activity started in 1995/1996 when Mayor Wellington Webb declared 1996 the year of the river. This generated a good deal of planning activity for more river-related projects both for park and hydrologic development. At the same time the USACOE was attempting to focus on urban water projects, starting with $13 to $15 million in funding and ending up with an $18 million price tag for two projects, one of those being the South Platte. Over seven years, Denver city leadership and the Colorado congressional delegation kept the South Platte in the funding mix. The Greenway Foundation, which evolved from the original development committee, was especially important in putting up the non-federal match to keep the project alive until 2006. Jeff Shoemaker (son of Joe Shoemaker) was appointed executive director of the Greenway Foundation. Key roles of this not-for-profit corporation included:

- Grant writing for river-related projects.
- Bridging river-related projects from one city or county administration to the next.
- Liaising between city and county agencies.
- Developing and running river-related programs such as educational, performing arts and tourism events; and
- Keeping real estate development professionals cognizant of the attributes of river-oriented development opportunities.

Jeff Shoemaker (personal communication, June 2008) feels that timing of effort is very important. Denver has had a number of strong mayors who have made river-related revitalization projects a high priority. Successful river recreation/access projects create demand for more of the same as well as connecting gaps in the river park/bicycle/pedestrian system. The current focus is on north/south river management plans and obtaining buy-in from the surrounding city and county for implementation. The focus has shifted from downtown Denver to surrounding communities and tributaries of the Platte River.

Management framework

The third part of the story was the creation of the Long Range Management Framework in 2000. Bar Chadwick of Mayor Wellington's South Platte River Commission chaired the planning effort. Her view (June 2008 interview) is that the framework helped consolidate much of the previous planning work, but with an eye to the future. The proposed South Platte River Commission was not created. But certain parts of the framework were implemented, including:

- A requirement that residential neighborhood associations within 200 feet of the river districts register with the city.
- The 150-foot setback from the river is used by all city planning agencies for project review.
- The greenway ordinance has been used; and
- Adaptive management is being utilized by the Denver Parks Department for more native species usage and for trail management.

Lessons learned

Key agencies and actors include the original Platte River Development Committee, which morphed into the Greenway Foundation; the City and County of Denver, which maintain parks and trail systems; certain city and county administrations that have pushed major river efforts; and state and federal legislators who have been key to funding projects. Although there was major community participation in decision-making in the early years, recent efforts have moved toward high-level public–private partnerships, with the Greenway Foundation playing multiple major roles. The physical transformation of the Platte River greenway park system, trail system and related private development is impressive. Key lessons include obtaining early physical results, keeping the river vision through multiple city/county administrations and key partnering negotiations to keep funding opportunities alive for extended periods of time. This is not a strong EJ case except from the point of view of providing green space access throughout the city, bicycle commuting access and economic development (residential) near the greenway.

Milwaukee River Greenway, Wisconsin, case study

The Milwaukee River has long been part of a downtown revitalization effort, but recent activity has focused on creating an upstream Milwaukee River Greenway plus water quality improvements throughout the Milwaukee River watershed. To obtain a sense of the character of the Milwaukee River area and its tributaries we highly recommend Eddee Daniels's (2008) book, *Urban Wilderness: Exploring a Metropolitan Watershed.*

Multiple organizations

Four organizations are involved with Milwaukee River restoration work: The Milwaukee Riverkeeper, which is primarily involved in water quality and habitat monitoring and improvements; the River Revitalization Foundation, which is involved with fundraising, property acquisition and management; the Urban Ecology Center, which runs educational programs for schoolchildren and other groups within the city; and the Milwaukee River Work Group, which is the coordinating group involved with Milwaukee River Greenway planning.

Milwaukee Riverkeeper: Milwaukee River basin partnerships began in the 1980s involving government, citizens, builders and other stakeholders in water quality and flooding issues. There are about 126 combined sewer overflows on the Milwaukee River, so storm event water quality and flooding are of significant concern. The Metropolitan Milwaukee Sewer District (MMSD) has a good water quality monitoring program but needs to break down data to a sub-watershed level. There is a need to implement stormwater phase II best management practices for Milwaukee and surrounding communities. The Milwaukee Riverkeeper's main mission is to provide educational and technical support to communities in the Milwaukee River watershed to implement water quality improvement programs. There have been some success stories, such as the Menomonee River Watershed Quality Initiative, the Kinnickinnic River Neighborhood Program and the Seminary Dam removal on Pigeon Creek, as well as production of water conservation education materials. The Milwaukee Riverkeeper is also part of the Southeastern Wisconsin Watershed Trust, which is a Chicago Wilderness-type plan for biodiversity and habitat protection throughout the region.

River Revitalization Foundation/Urban Ecology Center: The River Revitalization Foundation is also part of the Southeastern Wisconsin Watershed Trust, but at the present time it is focused on urban land trust activities. These activities include acquiring land and easements adjacent to the Milwaukee River, building connecting trails beyond city park areas along the river, invasive plant control and fundraising. The foundation is also starting a mini-grant program focused on small projects involving young schoolchildren. This activity ties in with the Urban Ecology Center, which uses the river as a nature interpretation and educational resource. The central location of the Center, close to the river, promotes communication.

Milwaukee River Work Group: The first coordinator of the Milwaukee River Work Group was Ann Brummitt, who came on board in spring 2006. The focus of the group is to develop a Milwaukee River Greenway Masterplan to preserve the wild aspect of the upstream Milwaukee River corridor while improving habitat, water quality and access. Existing city and county parks downstream interconnect with trails along the river. The Work Group proposes establishing a "viewshed" for the river corridor that would then be used to regulate new development through height control and setbacks. Its outreach through newsletters and meetings conveys the view that the upper Milwaukee River corridor is the "Central Park" of Milwaukee.

Lessons learned

The Milwaukee River Working Group exists as a sort of umbrella group, with its major scope being planning. The River Revitalization Foundation focuses on land acquisition and management and the Urban Ecology Center on educational programming. The Riverkeeper, which has been functioning longer, focuses on water quality monitoring and technical program implementation. These components seem to work well together and there is a network of community people who act as a kind of community think tank. For example, the Director of the Urban Ecology Center indicated that he helped initiate early discussions for the Greenway planning work and now gets involved when they need more "political horsepower." The existence of many city and county riverside parks helps lay the foundation for planning. There does seem to be potential conflict brewing about the idea of future height and setback controls for riverside development, especially high-rise apartment/condominium structures proposed along the upstream riverfront, which would be a distribution EJ issue of affordable housing versus open green space.

Tennessee River Greenway case study

The Tennessee Riverfront in Chattanooga, Tennessee, is also a legendary project in terms of urban riverfront revitalization. It encompasses several different organizations, including the River City Company, which plans and directs riverfront development projects with the City of Chattanooga; Outdoor Chattanooga, which orchestrates outdoor recreation activities and major events; The Trust for Public Land, which acquires land and easements within the Chattanooga waterfront and along tributaries; and the Tennessee River Gorge Trust, which acquires land and easements within the Tennessee River Gorge three miles south of the city.

From Mocosin Bend to the River City Company to master plan

Jim Bowen worked with Economic Development with the City of Chattanooga in 1982 (J. Bowen, personal communication, May 2008). At that time the city, Hamilton County and the State of Tennessee agencies, with funding from the Lyndhurst Foundation, started planning work on Mocosin Bend. This 1,000-acre area, just three miles south of the city, was rich in historical attributes and natural features. Kevin Lynch and Associates was charged with developing a plan for the area, which took three years and incorporated feedback from a local citizen task force. The final master plan was presented to an overflowing room of 1,400 citizens. In 1985, Bowen was hired with Lyndhurst Foundation funds (Coca-Cola Company) to implement the Mocosin Bend Plan and also inspect successful waterfront revitalization efforts in San Francisco, San Antonio and St. Paul, Minnesota. He felt that St. Paul had the best approach and adopted its form of bylaws and charter (J. Bowen, personal communication, May 2008).

By May 1986, $3.4 million in private funds had been raised. A board of directors of 12 individuals—six from government and six from the private sector—was appointed to create the River City Company, with a charter to improve the riverfront within the downtown area. The River City Company had no governmental power but did riverfront planning. From 1986 to 1988 Kevin Lynch and Associates were hired to do conceptual design and economic studies, and by 1992 there was a Tennessee River master plan. The purpose of the plan was to provide access to the riverfront through a connecting ribbon of public places with a continuous river walk. By May 1989 a 3-mile riverwalk plus one river park were in place. By 1992 the Tennessee Aquarium had been built. By 1993 the Walnut Street Pedestrian Bridge, the longest such bridge in the U.S., had been recommissioned. In 1999 Coolidge Landing Park became operational, and by 2005 most of the riverfront access park work was completed.

David Unruh, Director of Development with the River City Company (personal communication, May 2008), states that the organization's functions are to:

- Provide ongoing continuity, bridging administrations for city and county governments.
- Provide a neutral meeting place for private and public interests and mediate if needed.
- Stay focused on the same waterfront geographic area.
- Work as a technical advisory resource for downtown planning; and
- Recruit both private and public development partners for downtown.

Orchestrating major water events

Chattanooga Outdoors acts as a fundraiser/organizer for major events on the downtown waterfront. Prime examples are a national regatta, which the group lured from Atlanta, and the dragon boat races held twice a year. It also provides flat water paddling recreation information to residents and tourists.

Land preservation within the city

The Trust for Public Land (TPL) came to Chattanooga as the city was already working on the master plan. TPL's major role was acquiring land and easements to provide access to the Tennessee River tributaries and protect water quality and habitat. It often negotiates land sales or easements, and then transfers these to the city or county. The Trust also works with river and creekside landowners to be good stewards for land held as easements.

Tennessee River Gorge Land Trust

For land south of the city the major actor is the Tennessee River Gorge Land Trust (TRGLT). The group began 27 years ago as a citizen working group interested in protecting the gorge, which is 28 miles long, 2 miles wide and 2,000 feet deep and is one of the most scenic and natural stretches of the Tennessee River. The gorge contains roughly 27,500 acres rising from both sides of the river. Development pressure in the early 1980s spawned a formal committee, with assistance from TPL, which incorporated as a 501(c)(3) in 1986.

According to Jim Brown (personal communication, May 2008), the original TRGLT director, the group drew a line around the edge of the gorge using visibility from the water to the ridge. TRGLT then commissioned a detailed biological inventory of the whole 27,500 acres to determine which areas had unique or endemic ecology that should be preserved or was under threat. This work was used to set priorities for land acquisition. TRGLT now holds 6,500 acres in fee ownership, 700 acres leased from state agencies and some 3,000 to 4,000 acres in easements. TRGLT also has MOUs for stewardship management with the Tennessee Valley Authority (TVA), state forestry and other state lands. Many of those on its large board of directors are highly influential in the community. When asked about future challenges, Jim Brown volunteered that he was most concerned about:

- Succession and continuity of board members and reaching out to the new generation; and
- Funding for land acquisition and stewardship (although it is in good shape compared with other land trusts).

Lessons learned

This was a highly successful urban waterfront development program, although there wasn't much public participation throughout the process. The Greenway series of parks and walkways stretching from the TVA dam upstream is continuous down to the center city waterfront, where it moves to the opposite west shore with two new parks and connecting Walnut Street Pedestrian Bridge. TLP and TRGLT combine to acquire key land areas within and to the gorge area, respectively. A key element here was local capital from the Lyndhurst Foundation, which allowed early development of key planning efforts, plus the continuity of the River Company for urban areas and TLP and TRGLT for other areas. The ongoing promotion work by both the River Company for development and Chattanooga Outdoors for recreational tourism is outstanding. There does not seem to be much concern about water quality, but there is concern about potential water transfers outside the watershed for water-starved southern metropolitan areas. This is primarily an economic revitalization effort with not much social issue integration.

Summary and conclusions

This chapter reviews ways of supporting engagement in urban waterway revitalization. Historically there has been little theoretical or practical documentation of such efforts, especially with reference to urban waterways planning processes; however, there are now some emerging methods on engaging waterway communities and stakeholders in a positive process.

Social learning and co-production are two keys in this regard that have been utilized in both Europe and North America. In incorporating environmental justice issues such processes should also show interactive, distributive and deliberative justice.

This chapter also includes a review of highly varied organizational models with lesser and greater participation in decision-making. At one end one finds laudatory participatory efforts in the Bronx River, Onondaga Creek and Wildcat-San Pablo Creeks. For the South Platte, Guadalupe River and Tennessee River Greenway there are periods of participatory activity, but much is done by strong leaders/facilitators. The Milwaukee River Greenway is just starting so it is difficult to evaluate at this point.

There are also differences in focus, e.g., restoration vs. revitalization goals. Clearly the Guadalupe River in San Jose, the South Platte River in Denver and the Tennessee River in Chattanooga are mainly driven by economic development. The Wildcat-San Pablo Creek, Milwaukee River Greenway and to some degree Onondaga Creek are more focused on habitat preservation and natural river restoration. Social equity is a major issue for the communities involved with the Bronx River, the Milwaukee River and the Wildcat-San Pablo creeks in the sense that all creek communities receive benefits and none is disadvantaged.

Finally, there are different organizational schemas throughout all these projects, similar to what was seen in Chapter 4 (see Table 4.1). There are public–private partnerships, not-for-profits that bridge over time with different city or county administrations; urban land trusts and similar organizations that raise money, acquire land and/or easements and pass them on to public agencies; and groups that organize events and activities leading to more consciousness raising and use of urban waterways. There are organizations that perform many functions and some organizational schema that separate functions among many organizations. A potential research challenge is to more formally diagram and assess these organizational relationships and functions (Felleman, 1997; Vantainen & Marttunen, 2005).

References

Arnstein, S.R. (1969). A ladder of citizen participation. *Journal of the American Planning Association, 35*(4,) 216–224. doi: 10.1080/01944366908977225

Borowski, I., Le Bourhis, J., Pahl-Wostl, C., & Barraque, B. (2008). Spatial misfit in participatory river basin management: Effects on social learning. A comparative analysis of German and French case studies. *Ecology and Society, Resilience Alliance, 13*(1), Retrieved from www.ecologyandsociety.org/vol13/iss1/art7/

Bots, P.W.G., Bijlsma, R., Von Korff, Y., Van Der Fluit, N. & Wolters, H. (2011). Supporting the constructive use of existing hydrological models in participatory settings: A set of "rules of the game." *Ecology and Society, Resilience Alliance, 16*(2). Retrieved from www.ecologyandsociety.org/vol16/iss2/art16/

Costanza, R. & Ruth, M. (1998). Using dynamic modeling to scope environmental problems. *Environmental Management 22*(2), 183–195. doi: 10.1007/s002679900095

Daniel, E. (2008). *Urban Wilderness: Exploring a Metropolitan Watershed*. Chicago IL: The Center for American Places at Columbia College.

Daniell, K.A., White, I., Ferrand, N., Ribarova, I.S., Coad, P., Rougier, J-E., Hare, M., Jones, N.A., Popova, A., Rollin, D., Perez, P. & Burn, S. (2010). Co-engineering participatory water management processes: Theory and insights from Australian and Bulgarian interventions. *Ecology and Society 15*(4). Retrieved from www.ecologyandsociety.org/vol15/iss4/art11/

Eden, S. & Tunstall, S.M. (2006). Ecological versus social restoration? How urban river restoration challenges but also fails to challenge the science-policy nexus in the United Kingdom. *Environment and Planning C: Politics and Space, 24,* 661–680. doi: 10.1068/c0608j

Eden, S., Tunstall, S.M. & Tapsell, S.M. (2000). Translating nature: River restoration as nature-culture. *Environment and Planning D: Society and Space, 18,* 257–273. doi: 10.1177/026377580001800101

Enserink, B., Patel, M., Kranz, N. & Maestu, J. (2007). Cultural factors as co-determinants of participation in river basin management. *Ecology and Society 12*(2). Retrieved from www.ecologyandsociety.org/vol12/iss2/art24/

Environment Agency (EA). (2013). *Tame, Anker and Mease Catchment Partnership: Catchment Management Plan*. Coventry, U.K.: Environment Agency, The Wildlife Trust for Birmingham and the Black County, Trent Rivers Trust and Warwickshire Wildlife Trust.

Felleman, J. (1997). *Deep Information: The Role of Information Policy in Environmental Sustainability*. Greenwich, CT: Ablex.

Firehock, K., Flanigan, F. & Devlin, P. (2002). *Community Watershed Forums: A Planner's Guide*. Baltimore, MD: Alliance for the Chesapeake Bay.

Hagerman, C. (2007). Shaping urban neighborhoods and nature: Urban political ecologies of urban waterfront transformations in Portland, Oregon. *Cities, 24*(4), 285–297. doi: 10.1016/j.cities.2006.12.003

Holt, A. R., Moug, P., & Lerner, D.N. (2012). The network governance of urban river corridors. *Ecology and Society, 17*(4), 25–39. doi: 10.5751/ES-05200-170425

Hopkins, A.W. (2005). *Groundswell: Stories of Saving Places, Finding Community*. San Francisco, CA: The Trust for Public Land.

Huitema, D., Cornelisse, C. & Ottow, B. (2010). Is the jury still out? Toward greater insight in policy learning in participatory decision processes—the case of Dutch citizens' juries on water management in the Rhine Basin. *Ecology and Society, 15*(1). Retrieved from www.ecologyandsociety.org/vol15/iss1/art16/

Innes, J. & Booher, D. (2000). Reframing public participation strategies for the 21st century. *Planning Theory and Practice, 5,* 419–436. Retrieved from https://escholarship.org/uc/item/4gr9b2v5

Kuper, M., Dionnet, M., Hammani, A., Bekkar, Y., Garin, P. & Bluemling, B. (2009). Supporting the shift from state water to community water: Lessons from a social learning approach to designing joint irrigation projects in Morocco. *Ecology and Society, 14*(1). Retrieved from www.ecologyandsociety.org/vol14/iss1/art19/

Lamers, M., Ottow, B., Francois, G. & Von Korff, Y. (2010). Beyond dry feet? Experiences from a participatory water-management planning case in the Netherlands. *Ecology and Society, 15*(1). Retrieved from www.ecologyandsociety.org/vol15/iss1/art14/

Lawrence, R.L., Daniels, S.E. & Stankey, G.H. (1997). Procedural justice and public involvement in natural resource decision making. *Society & Natural Resources, 10*(6), 577–589. doi: 10.1080/08941929709381054

Leach, W.D. & Pelkey, N.W. (2001). Making watershed partnerships work: a review of the empirical literature. *Journal of Water Resources Planning and Management* 127(6), 378–385. doi: 10.1061/(ASCE)0733-9496(2001)127:6(378)

Loucks, D.P., Kindler, J. & Fedra, K. (1985). Interactive water resources modeling and model use: An overview. *Water Resources Research*, 21(2), 95–102. doi: 10.1029/WR021i002p00095

Marks, J.S. (2004). Negotiating change in urban water management: Attending to community trust in the process. *Proceedings of WSUD 2004: Cities as Catchments: International Conference on Water Sensitive Urban Design* A.C.T.: Engineers Australia (pp. 203–215). Retrieved from https://search.informit.com.au/document Summary;dn=767146301410701;res=IELENG;subject=HistoryMcGinnis, M.V., Wooley, J. & Gamman, J. (1999). Forum: biological conflict resolution: Rebuilding community in watershed planning and organizing. *Environmental Management*, 24(1), 1–12. doi: 10.1007/s002679900210

McGowan, C. P., Lyons, J.E., Smith, D. R. (2015). Developing objectives with multiple stakeholders: Adaptive management of Horseshoe crabs and red knots in the Delaware Bay. *Environmental Management* 55, 972–982, doi 10.1007/s00267-014-0422-8

Manno, J., Smardon, R., De Pinto, J., Cloyd, E.T. & Del Granado, S. (2008). The Use of Models in Great Lakes Decision Making: An Interdisciplinary Synthesis. Randolph G. Pack Environmental Institute, SUNY/ESF. Occasional paper 16 Retrieved from www.esf.edu/es/documents/GreatLakesRpt.pdf

Méndez, P.F., Isendahl, N., Amezaga, J.M. & Santamaría, L. (2012). Facilitating transitional processes in rigid institutional regimes for water management and wetland conservation: Experience from the Guadalquivir Estuary. *Ecology and Society*, 17(1). doi: 10.5751/ES-04494-170126

Moellenkamp, S., Lamers, M., Huesmann, C., Rotter, S., Pahl-Wostl, C., ... Pohl, W. (2010). Informal participatory platforms for adaptive management. Insights into niche-finding, collaborative design and outcomes from a participatory process in the Rhine Basin. *Ecology and Society*, 15(4), 41–62. Retrieved from www.ecologyandsociety.org/vol15/iss4/art41/

Moran, S. (2010.) Cities, creeks, and erasure: Stream restoration and environmental justice. *Environmental Justice*, 3(2), 61–69. doi: 10.1089/env.2009.0036

Moran, S. (2007). Stream restoration projects: A critical analysis of urban greening. *Local Environment*, 12(2), 111–128.

Mostert, E., Pahl-Wostl, C., Rees, Y., Searle, B., Tàbara, D. & Tippett, J. (2007). Social learning in European river-basin management: Barriers and fostering mechanisms from 10 river basins. *Ecology and Society*, 12(1). Retrieved from www.ecologyandsociety.org/vol12/iss1/art19

Onondaga Environmental Institute (OEI). (2008). *Case Studies Guide: Conceptual Alternatives for Onondaga Creek*. Syracuse, NY: Onondaga Environmental Institute. Retrieved from www.oei2.org/OEIResources_CaseStudiesGuide.html

Onondaga Environmental Institute (OEI). (2009). *Onondaga Creek Conceptual Revitalization Plan*. Syracuse, NY: Onondaga Environmental Institute. Retrieved from www.oei2.org/OEIResources_OCRPDRAFT.html

Otto, B., McCormick, K. & Leccese, M. (2004). *Ecological Riverfront Design: Restoring Rivers, Connecting Communities*. Chicago, IL: American Planning Association Advisory Service Report Number 518–519.

Pahl-Wostl, C., & Hare, M. (2004). Process of social learning in integrated resources management. *Journal of Community & Applied Social Psychology, 14*, 193–205. doi: 10.1002/casp.774

Pahl-Wostl, C., Craps, M., Dewulf, A., Mostert, E., Tàbara, D. & Taillieu, T. (2007). Social learning and water resources management. *Ecology and Society, 12*(2). Retrieved from www.ecologyandsociety.org/vol12/iss2/art5/

Pahl-Wostl, C., Mostert, E. & Tàbara, D. (2008). The growing importance of social learning in water resources management and sustainability science. *Ecology and Society, 13*(1). Retrieved from www.ecologyandsociety.org/vol13/iss1/art24/

Perreault, T., Wraight, S. & Perreault, M. (2012). Environmental injustice in the Onondaga Lake waterscape, New York State, USA. *Water Alternatives, 5*(4), 485–506.

Piegay, H., Dupont P. & Faby J.A. (2002). Questions of water resources management. Feedback on the implementation of French SAGE and SDAGE plans (1992–2001). *Water Policy 4*(3), 239–262. doi: 10.1016/S1366-7017(02)00008-9

Petts, J. (2006). Managing public engagement to optimize learning: Reflections from urban river restoration. *Human Ecology Review, 13*, 172–181.

Petts, J. (2007). Learning from learning: Lessons from public engagement and deliberation in urban river restoration. *The Geographical Journal, 173*(4), 300–311.

Platt, R.H. (2006). Urban watershed management sustainability: One stream at a time. *Environment, 48*(4), 26–42. doi: 10.3200/ENVT.48.4.26-42

Priscolli, J.D. (2004). What is participation in water resources management and why is it important? *Water International, 29*(2), 1–7. doi: 10.1080/02508060408691771

Riley, A.L. (1989). Overcoming federal water policies: The Wildcat-San Pablo Creeks Case. *Environment, 31*(10), 12–31. doi: 10.1080/00139157.1989.9928987

Riley, A.L. (1998). *Restoring Streams in Cities: A Guide for Planners, Policymakers and Citizens.* Washington, D.C.: Island Press.

Riley, A.L. (2016). *Restoring Neighborhood Streams: Planning Design and Construction.* Washington, D.C.: Island Press.

Rowe, G., & Frewer, L.J. (2000). Public participation methods: A framework for evaluation. *Science Technology & Human Values, 25*(3), 3–29. doi: 10.1177/016224390002500101

Rowe, G., & Frewer, L.J. (2004). Evaluating public participation exercises. *Science Technology & Human Values, 29*(4), 512–556. doi: 10.1177/0162243903259197

Runfola A-M & Weiss, J. (2007). *Bronx River Classroom: The Inside Track for Educators.* Bronx, NY: Bronx River Alliance.

Sabatier, P.A., Focht, W., Lubell, M., Trachtenberg, Z., Vedlitz, A. & Matlock, M. (2005). *Swimming Upstream: Collaborative Approaches to Watershed Management.* Cambridge MA: The MIT Press.

Scholz, G., VanLaarhoven, J., Phipps, L., Favier, D. & Rixon, S. (2002). Managing for river health: Integrating watercourse management, environmental water requirements and community participation. *Water Science and Technology 45*, 209–213.

Selman, P., Carter, C., Lawrence, A. & Morgan, C. (2010). Re-connecting with a neglected river through imaginative engagement. *Ecology and Society, 15*(3). Retrieved from www.ecologyandsociety.org/vol15/iss3/art18/

Shoemaker, J. (1981). *Returning the River to the People.* Westminster, CO: Tumbleweed Press.

Silverman, D. (2003). Analyzing talk and text. In N.K. Denzin & Y.S. Lincoln (Eds.), *Collecting and Interpreting Qualitative Methods* (pp. 340–317), Thousand Oaks, CA: Sage Publications.

Smardon R.C., Felleman, J.P. & Senecah, S.L. (1996). *Protecting Floodplain Resources: A Guidebook for Communities*. Washington, D.C.: Prepared for the Federal Interagency Floodplain Management Task Force, FEMA publication 268 U.S. Gov. Printing Office.

Smolko, B., Huberd, R. & Tam-Davis, N. (2002). Creating meaningful stakeholder involvement in watershed planning in Pierce County, Washington. *Journal of the American Water Resources Association*, 38, 981–994. doi: 10.1111/j.1752-1688.2002.tb05539.x

Spirn, A.W. (2005). Restoring Mill Creek: Landscape literacy, environmental justice and city planning and design. *Landscape Research*, 30(3), 395–413. doi: 10.1080/01426390500171193

Strauss, A. (1987). *Qualitative Analysis for Social Scientists*. New York, NY: Cambridge University Press.

Syme, G.J. & Nancarrow, B.E. (2002). Evaluation of public involvement programs: Measuring justice and process criteria. *Water*, 29(4), 18–24.

U.S. Department of Interior National Park Service (USDI-NPS). (1988). Riverwork Book. Washington D.C.: USDI, National Park Service, Mid-Atlantic Planning Office.

U.S. Environmental Protection Agency (USEPA). (2002). *Community Culture and the Environment: A Guide to Understanding Sense of Place*. EPA 842-B-01-003. Washington, D.C.: USEPA Office of Water.

U.S. Environmental Protection Agency (USEPA). (2003). *Getting in Step: A Guide to Watershed Campaigns*. EPA 841-B-03-002. Washington, D.C.: USEPA Office of Water

Vantainen, A., & Marttunen, M. (2005). Public involvement in multi-objective water level regulation development projects—evaluating the applicability of public involvement methods. *Environmental Impact Assessment Review*, 25, 281–304. doi: 10.1016/j.eiar.2004.09.004

Von Korff, Y., D'Aquino, P., Daniell, K.A. & Bijlsma, R. (2010). Designing participation processes for water management and beyond. *Ecology and Society 15*(3). Retrieved from www.ecologyandsociety.org/vol15/iss3/art1/

Von Korff, Y., Daniell, K.A., Moellenkamp, S., Bots, P. & Bijlsma, R.M. (2012). Implementing participatory water management: Recent advances in theory, practice, and evaluation. *Ecology and Society*, 17(1). doi: 10.5751/ES-04733-170130

Wessells, A.T. (2010). Place based conservation and urban waterways: Watershed activism in the bottom of the basin. *Natural Resources Journal*, 50, 539–557. Retrieved from http://digitalrepository.unm.edu/nrj/vol50/iss2/14

Wondolleck, J.M. & Yaffee, S.L. (2000). *Making Collaboration Work: Lessons from Innovation in Natural Resource Management*. Washington, D.C: Island Press.

Yocom, K. (2014). Building watershed narratives: An approach for broadening the scope of success in urban stream restoration. *Landscape Research*, 39(6), 698–714. doi: 10.1080/01426397.2014.947249

Zorrilla, P., Carmona, G., De la Hera, Á., Varela-Ortega, C., Martínez-Santos, … Henriksen, H.J. (2010). Evaluation of Bayesian networks as a tool for participatory water resources management: Application to the upper Guadiana basin in Spain. *Ecology and Society 15*(3). Retrieved from www.ecologyandsociety.org/vol15/iss3/art12/

6 Restoring streams and relationships

Richard Smardon, Sharon Moran, Jill Weiss and Blake Neumann

Introduction

From the opening chapter of this book, it was noted that many communities have been disconnected from urban waterways, are burdened with environmental justice (EJ) issues, and in some cases have negative views of urban waterways (Schauman & Salisbury, 1998), as studied in the Puget Sound region in Washington State. On the other hand, there are sometimes rich narratives about urban waterways, such as those for Onondaga Creek, told by both European descendants and Native Americans (Moran et al., 2016; Perreault et al., 2012). A key question for the future will be how to reconnect these place-based narratives to address environmental and social grievances and incorporate them into urban waterway revitalization. This chapter briefly reviews appropriate literature, presents some case studies and explores the roles elders can play in restoring urban waterways.

Review of the issues: Local knowledge to engagement

With his work on Hunter Valley in New South Wales, Australia, Hillman (2004, 2005) stresses the need for rediscovery of the historical and geographic context. In his work he strives to bring to the surface the colonial history of river management, the historical inclusion or exclusion of stakeholders in decision-making and the narrow definition of affected community. Spirn (2005) addresses the need for local landscape literacy to address environmental justice issues with the Mill Creek Neighborhood in Philadelphia, Pennsylvania. Landscape literacy in the Mill Creek case means knowing the environmental history of Mill Creek as well as the community along the creek. Rogge et al. (2005) stress understanding EJ issues as major drivers for action in the Chattanooga Creek case in Tennessee. This case involves a minority community subject to both flooding and water pollution along the creek. These examples underline the need to understand the historical community context and knowledge base, especially when there have been environmental justice issues.

Nassauer et al. (2001) note how influential cultural values—attachments to the riparian landscape that are more social or symbolic than physical—are for the riparian landscape, and how attention to such values supports the public acceptance of ecologically innovative design. Likewise, May (2006) calls for attention to understanding the cognitive connectivity in riparian landscapes. In many cases, there has been a distinct difference between a local community's desired revitalization objectives (public health and aesthetics) and those of the experts (e.g., hydrologic engineering and ecological). When such differences exist, it's important to understand them and then proceed with an inclusive revitalization process. May (2006) proposes a process that includes a dialogue between parties to bring out key objectives and then working collaboratively through the revitalization process.

Petts (2006, 2007) explains how local knowledge can lead to shared expert and local learning and understanding with waterway restoration projects in the U.K. Selman et al. (2010) propose use of imaginative engagement of "catchment consciousness" to push citizens to invest time in researching river history and geography for the same objective. An example would be imagining a different waterway future given its history.

Wessels (2010) proposes building capacity to address environmental equity as part of urban watershed management via experimental symbolic, identity-based means of social engagement. Both Wessels (2010) and Wolch (2007) support place-based activism as a means of social engagement for the Los Angeles River in California. In the Los Angeles River case, there have been locally generated proposals for naturalization of the river plus river events and tour guides to re-engage residents with the river.

Brody, Highfield, & Peck (2005) have used surveys and network analysis to uncover environmental perceptions related to spatially dependent access to the river landscape in San Antonio, Texas. This work has highlighted geographic networks of issue-based activism and localized hot spots for San Antonio River neighborhoods. Some of these hot spots identified involve minority open space needs, which could be considered an environmental justice issue because of lack of access. There is a need to identify community concerns regarding urban waterways, so they can become part of the revitalization dialogue.

Wade et al. (2002) propose the use of social analysis and place-based evaluations to assist communities to formulate environmental visions for decision-making and translate them into specific stream naturalization strategies for the Chicago area. Likewise, Beierle & Konisky (2000) stress introducing public values into governmental agency decision-making to resolve conflict and build trust in agencies. Petts (2006, 2007) documents the use of gatekeepers of knowledge, interest and values as part of a privileged narrative, which is critical to instrumental and communicative learning. It is crucial to emphasize local knowledge, concerns and perceptions, and there is a need to continue processing them through analysis and social learning to decision-making.

Finally, Moran (2007, 2010) calls for socially and culturally sensitive metrics to evaluate waterway restoration projects versus technical and "managerialist" construction of the problem definition. Such metrics would need before-and-after tracking of public evaluations of waterway revitalization/naturalization projects. A good example of such evaluation is the work of Tunstall et al. (1999) on the stability of public responses to urban waterway restoration before and after implementation on the River Skerne in Darlington, U.K. As we will see from the River Skerne case study, pre- and post-project evaluation is critical to judge project success and acceptability of outcomes.

The following sections comprise three case studies addressing connectivity issues: Mill Creek, Philadelphia (Spirn, 2005); the South Bronx River Greenway; and the River Skerne in Darlington, U.K. All three cases involve elements of connectivity back to urban waterways. The Mill Creek and South Bronx cases involve strong environmental justice issues; all three cases illustrate mechanisms for community engagement.

Mill River case study: Building creek landscape literacy

Anne Spirn (2005) worked with Mill Creek for more than four years (1987–1991) with her West Philadelphia Landscape Project, and she has documented this case study in her 2005 article. As part of this project, her students created a digital database covering the demographic and physical features of the area. She worked with several parties (including her students at the University of Pennsylvania and the Philadelphia City Planning Department) to bring attention to the flooding, subsidence, water pollution and environmental justice issues of this urban creek neighborhood. Later, she and her students worked with the Sulzberger Middle School to develop landscape literacy for the Mill Creek area. Some of these children did not know there was a creek nearby; nor did they have much sense of the history of their neighborhood. They helped develop a historical timeline from the 17th century forward. Other components of the literacy program included small learning communities and development of proposals for how the creek could be transformed from a liability to an asset.

The result of this project put Mill Creek "on the map" in terms of local, national and international recognition. According to Spirn, "from 1996 to 1999, there were over a million visits to the West Philadelphia Landscape Project website, from more than 90 countries on six continents" (Spirn, 2005, p. 407). While the project opened doors for local collaborations for creek-related projects, Spirn also noted that some of the needed creek demonstration projects did not go forward with city agencies.

Spirn noted that she thought the worst effect of environmental literacy was highlighting the physical hazards to health and safety, and she continued to point out that an even greater injustice was "the internalization of shame in one's neighborhood" (Spirn, 2005, p. 409). In other words, environmental injustices were reinforced as the children learned more about the environmental and community historical context. On the other hand, Spirn noted that studying the creek's natural and built features brought the place alive for the middle-school students, in turn opening wider vistas that introduced them to broader social, political and environmental issues. Interestingly, she also argued that planning professionals were not as landscape literate as the middle school students and local residents.

South Bronx River Greenway

The Bronx River is the only freshwater river in New York City. It runs from affluent Westchester County through the New York Botanical Garden and the Bronx Zoo down to blighted areas of the Bronx and into the East River. Centuries of industrial boom cycles and misguided city planning have dramatically changed the physical nature of the river and Bronx residents' view of it (Byron, 2004; Perini & Sabbion, 2017). By the 1970s, the river had reached its lowest point, with active trash and vehicle dumping, high crime rates in adjacent parks and vast algal blooms. This moment coincided with a growing environmental movement, and in 1974 Bronx Borough Police Commander Anthony Bouza organized community meetings to address the river's social and ecological health (Taylor, 2007). In response to increased community interest, Ruth Anderberg organized cleanups in the West Farms area of the river, and the Bronx River Restoration Project (BRRP) was born (Runfola & Weiss, 2007).

The cleanups continued for decades and involved many individual volunteers and in-kind services from organizations. In one case, the National Guard provided personnel and equipment to remove abandoned cars from the river (DeVillo, 2015). In another, Juanita "Ma" Carter, an active West Farms community member for 40 years, started the "Not In My Neighborhood You Don't" events. These were part street celebration and part demonstration to push crime out of the community through symbolic action and ownership such as park adoption. Ma Carter would later be awarded a Park Warden title for her work at River Park (Bronx River Alliance, 2017).

These activities prompted high levels of neighborhood engagement. Anderberg also formed the Youth Conservation Corps around this time, which spearheaded the Bronx community garden movement that endures today. They grow organic vegetables, provide food for the needy and educate children and the public about the necessity of greenery for community development. Such programs and organizations include the Bronx GreenUp Program at the New York Botanical Garden and Greenthumb, a program of the City Parks Department.

These demonstrations of ownership and interest surely led to the Bronx River's designation as an American Heritage River. In 1980, the Bronx River

Master Plan was created. Programming and cleanup activities continued along the river by an array of organizations (Hopkins, 2005; Marshall, 2001).

In 1986, Nancy Wallace took over leadership of BRRP and engaged New York City Parks Commissioner Henry Stern and Mayor Ed Koch to commit funds for ecological restoration, with some small successes. By 1990, however, it was clear that community work and small grants might not be enough to clean up the river.

New York City has a combined sewer system where street runoff and household sewage share the same system. During heavy rain or snow-melt, NYC's sewage treatment facilities (14 in total across the city) become overloaded. When this happens, the wastewater is diverted without treatment to the nearest creek, river or bay. This diverted liquid is known as combined sewer overflow or CSO, and the diversions are known as CSO events (Runfola & Weiss, 2007). CSOs were and continue to be a problem that seems to be out of the hands of the people.

CSOs are a public health concern because of the unknown pollutants picked up in the street runoff. More importantly, "they are the biggest source of disease-causing pathogens (e.g., fecal coliform bacteria) in the waters surrounding NYC" (Runfola & Weiss, 2007, p. 148). There are five discharge points located on the Bronx River in the Bronx, with an annual discharge of around 558 million gallons (Runfola & Weiss, 2007). The NYC Department of Environmental Protection (DEP) is responsible for managing water quality in NYC.

Once the central problem was clearly defined, engagement with the New York City DEP was imperative, and the Bronx Council for Environmental Quality made CSOs and the health of the river a priority. In 1997 Jenny Hoffner convened 60 organizations and asked the Bronx River Working Group to pull together varied restoration and park plans into a cohesive Greenway Plan and River Restoration Plan that considers the corridor as an entire system.

With monetary help from the USDA Urban Resources Partnership Program, the U.S. Department of Transportation and the EPA Wetlands Protection Program, the Bronx River Working Group brought together community groups, nonprofit organizations, businesses and others. Many projects have led to significant improvements, including improving water quality, habitat restoration and increased public access to the watersheds. These have been achieved through public outreach projects such as the Adopt the River Program, thereby gaining community involvement—a crucial part of significant river cleanup. Partnering groups include:

- The Point (often presented as POINT) Community Development Corporation, which played a part in the revitalization of the river and the creation of park space in the Hunts Point neighborhood. Established in 1998, the organization acted as a community center for the neighborhood and a way for citizens to become more political in Hunts Point.

- Majora Carter's nonprofit Sustainable South Bronx was founded out of the Bronx River Alliance as well, becoming the premier green job training and economic development organization in the South Bronx. Sustainable South Bronx established the Bronx Environmental Stewardship Training Program to cultivate talent and train participants for practical green jobs.
- In 1998, Adam Greene founded the nonprofit Rocking the Boat with the goal of developing the talent of local high school youth and teaching them boatbuilding skills. This program is successful both in the connections it makes with participants' academic work and in the community connections it strengthens.
- Youth Ministries for Peace and Justice was another nonprofit established by Alexie Torres-Fleming. This group partnered with the Bronx River Alliance and other organizations to transform a concrete plant into a scenic park that preserves the industrial infrastructure of the area's past.
- Additional stakeholder groups include the U.S. Army Corps of Engineers, the New York State Department of Environmental Conservation, the Westchester County Department of Environmental Planning, the New York City Restoration Project and the Fordham University Community Service Program.

In 1999, with the help of Henry Stern of NYC Parks and Jenny Hoffner of the Bronx River Working Group, a vision of the Bronx River Greenway was formed. The project was estimated to cost $60 million and be finished by 2010. These estimates ultimately proved inaccurate.

In 2000, the first Amazing Bronx River Flotilla was held by the Bronx River Restoration Project, ushering in an era of exciting recreation opportunities. In 2001 the Bronx River Alliance was formed from the Working Group. Having overseen the development of the Greenway and Restoration Plan, Hoffner stepped down as the new organization emerged.

The Bronx River Alliance's organizational structure and governance built upon the Working Group's success by actively involving a diverse base of stakeholders in defining and carrying out its mission. The Alliance's organizational and individual partners belong to one or more teams, each of which is responsible for an aspect of Alliance programming. The team structure integrates the involvement of Alliance partners with the day-to-day work of the professional staff. The structure also ensures that values of representation and inclusion are carried into the implementation of programs. The teams bring together community residents and organizations, educators, scientists and other professionals, as well as representatives of city, state and federal agencies.

The Alliance staff is headed by an executive director, who also serves as the park's Bronx River administrator. Like other public–private partnerships, the alliance operates as a not-for-profit organization, and also receives significant in-kind support from New York City Parks. The Alliance staff includes coordinators for each of the four major program areas (ecology, education, greenway and outreach), as well as the Bronx River Conservation Crew, a full-time workforce that maintains and monitors the Bronx data recorder and carries out restoration projects in conjunction with park staff.

The greenway coordinator is the Alliance's principal liaison to the public agencies implementing the many capital projects that make up the Bronx River Greenway. The $120 million in public funding commitments the Alliance and its partners have secured since 1999 is allocated to these agencies—chiefly NYC Parks and New York State Department of Transportation (NYSDOT)—and budgeted to the individual projects.

The Bronx River Alliance has become a community force, but still must cooperate with one branch of New York City government, NYC Parks, while opposing the sewage policy of the NYC Department of Environmental Protection. In 2016 the Bronx River Alliance served 1,000 volunteers who provided 3,000 hours, planted 2,746 trees, collected 741 full trash bags and had 6,466 participants at their events.

The stakeholders haven't always agreed about revitalization goals and objectives, as noted by several scholars (Campbell, 2006; Marshall, 2001). Marshall particularly notes the opposition from some to spending money on environmental restoration of the waterway given other important environmental projects in the New York City area. Another conflict takes the perspective that a poor and minority community has suffered an environmental injustice due to polluted water and lack of access to open space and recreation opportunities, and thus needs resources to remedy this EJ situation.

On a positive note, many environmental activist groups are involved in the Bronx River Alliance, addressing public outreach about such environmental concerns as wildlife conservation, environmental justice, water quality and recycling. With education comes the linkage to involve many community residents through these organizations. However, there is considerable debate as to their overall long-term success. Marshall (2001) and Hopkins (2005) argue that while small nonprofit organizations may not be triggering huge legislative changes, they do succeed in making people more aware of the current situation of our environment. Furthermore, they get people in the community involved, which provides hope for future generations.

No less significant than the funding commitments are the commitments by the Bronx River Alliance to implement and prioritize the greenway's design and construction and to work collaboratively to move the greenway forward in a way that reflects the unified vision of a uniquely diverse body of stakeholders. The challenges in realizing the greenway include the realities of limited staffing, competing priorities and genuine differences among public agency standards, protocols and cultures.

The Bronx River Alliance embodies its members' determination to overcome those challenges by facilitating, coordinating, problem solving and spurring its partners into action. Thus, the Bronx River Corridor

Restoration and Greenway is an umbrella-like network of groups, which in turn reconnects to neighborhoods through myriad groups and activities.

Skerne River restoration

Tunstall et al. (1999, 2000) conducted a pre- and post-survey of three river projects in the U.K. to assess attitudes toward the designs as implemented, perceived cost-effectiveness and public satisfaction with the level of consultation. Only the River Skerne project is covered here, as it is the most urban river restoration project.

The River Skerne project is a 2-kilometer section running through Darlington Borough Council parkland (also see Chapter 2). The river had been radically altered due to industrialization, residential development and flood protection works. Much of the floodplain had been eliminated and the river channel had become straight, featureless and overgrown. The river was subject to some water pollution but had improved slowly. The river naturalization project objective was to restore the river to a more natural condition by creating meanders, backwaters and wetlands, and through utilization of bio revetments with live vegetation. In other parts, deflectors were used to introduce flow sinuosity. Banks were re-profiled to a shallower slope to encourage aquatic plants. Fifteen visually intrusive outfalls were replaced with underwater outfalls, to improve appearance and decrease the impacts of pollution.

The Darlington residents who were surveyed were mostly owner-occupiers and lived in urban/industrial areas close to the town center. They were likely to live near the restored river on riverside property with an outlook across local-authority parkland, or within 500 meters of the river. Most (81 percent) Darlington residents visited the river within the previous 12 months, with 42 percent visiting weekly during the winter (Tunstall et al., 2000, p. 366). Walking was the main activity, but wildlife viewing also attracted significant numbers of visitors. Many residents surveyed before and after project implementation cited wildlife and school-related activities as their reason for visiting the river. Darlington respondents believed the flood risk to be lower after restoration because the river was wider and clearer, plus the project enhanced meanders and introduced wetlands. Increased recreational access via footpaths and bridges was appreciated by the Darlington residents, and the change was dramatic because of the previous degraded state of the river.

There were extensive public consultation processes for the Skerne project. While 73 percent of Darlington residents were involved in some way, only 16 percent were involved with interactive consultation (Tunstall et al., 2000, p. 369). Most felt they were not involved with decision-making for the project. Still, local people's views were also sought on a proposed all-weather path and footbridge through a pre-scheme survey, and these were incorporated into recreation and amenity-related design components.

Case studies summary

From these various case studies, it is possible to draw a number of lessons learned regarding reconnecting local communities with urban water bodies in need of revitalization:

- The first lesson is the value of reconstructing the historical context of the urban waterway itself and its relationship to the surrounding community, as illustrated by Mill Creek in Philadelphia (Spirn, 2005) and Hunter Valley in New South Wales, Australia (Hillman, 2004, 2005);
- The second lesson is to build a mechanism for waterway stakeholders to be engaged in an ongoing dialogue that promotes social learning, as illustrated by the South Bronx (Marshall, 2001), the Los Angeles River (Wessels, 2010; Wolch, 2007) and work in the U.K. (Petts, 2006, 2007);
- Third, to understand and address the fact that the revitalization objectives of waterway residents may not necessarily agree with those of government agencies or waterway revitalization experts (Nassauer et al., 2001; Petts, 2006, 2007); and
- Fourth, to devise ways of developing socially and culturally sensitive metrics that can be used to assess pre- and post-project outcomes, as illustrated by the River Skerne project (Tunstall et al., 1999, 2000).

Elders, restoration and justice

Many examples suggest that senior citizens (elders) can play a special role in stream restoration projects. This is not to claim that there are necessarily more elders doing this work, or that their participation levels are higher than those of other demographics, but rather that their role(s) in restoration may be distinct in important ways that are not frequently acknowledged. To help unpack the complex identities of older adults working in this realm, this essay explores the role(s) of elders in environmentalism, including their contributions to activism generally. Intersections among categories of other identities (race and ethnicity) will also be touched upon briefly. For some elders, it appears that working on restoration initiatives is more than simple voluntarism; it is rather an act that has specific significance relative to values, ethics, justice and memory.

Aging background

For some time, aging has been perceived as a descent into obsolescence. According to some, old age is considered to extend from approximately age 65 until the end of life (Achenbaum, 1978). In American culture today, the associations with the terms "senior citizen" and "elder" are often negative. These negative connotations aren't new or uniquely American; they have been present in many cultures throughout history. Biblical accounts have represented the elderly as weak and socially rejected; authors from classical antiquity also expressed this sentiment. The Roman philosopher Seneca (4 BC–AD 65) referred to old age as a "disease" (Achenbaum, 2010); Juvenal (AD 60–130), a Roman satirist, provided a vivid description of the physical deterioration of aging, describing elders as "hideous" and "deformed" and asserting that a long life concludes with "everlasting sadness" (de Beauvoir, 1972, pp. 121–122). In the modern period as well, aging has often been represented as primarily a period of decrepitude; for example, Elie Metchnikoff refers to aging as "an infectious, chronic disease, which is manifested by a degeneration, or an enfeebling of the noble elements," (1905, p. 48). In the 1970s, Bernice Neugarten expressed a slightly more nuanced view of aging, differentiating between two groups of seniors: the so-called "young old" and the "old-old," representing an "age-old distinction made between a 'green old age' and 'decrepitude'" (Achenbaum, 2010).

The term "ageism" refers to a pejorative categorization of older people, specifically a form of discrimination against individuals based on their age alone. One strain of ageism emerged during the Industrial Revolution and the machine age, and relates to the implications of the type of practical knowledge held by elders. At that time, the new industrialists began expressing a preference for more youthful employees to work with the efficient new machines, threatening the employment of older craftspeople (Lynd & Lynd, 1929, pp. 42–43). This eventually led to what is referred to by Achenbaum as an "[unmistakable] downward trend in old-age employment" (Schaie & Achenbaum, 1993; Achenbaum, 1978). Beyond the context of employment in factories, similar patterns of age-based discrimination have also been found in other more socially oriented institutions, including faith-based communities and health care centers (Achenbaum, 2005). Yet another strain of ageism casts seniors as "greedy" in that their consumption of resources is displacing what would otherwise be saved for future generations (Quadagno, 1989). This type of ageism has a decidedly utilitarian quality, suggesting that there is some deplorable sort of inefficiency associated with living a longer life.

With the increase in life expectancy of the past century alone, a re-evaluation of contributions made by seniors in environmentalism overall is well overdue. Surprisingly, the literature on this question is sparse; the topic has been addressed in some news stories (Moyer, 2015; Wayman, 2015) and some academic papers (Frumkin et al., 2012; Pillemer et al., 2017), but more comprehensive studies would continue exploring the role of seniors in environmental activism and conservation, documenting their participation in and contributions to initiatives. Still, there is no shortage of demographic data and projections, explaining that populations are aging and diversifying (e.g., Hogan et al., 2008; Ortman et al., 2014). Meanwhile, a preliminary exploration reveals a trend of active volunteer engagement in a wide range of tasks related to stream restoration, and this comes in addition to greater action in the realm of environmental issues overall.

One important theme is self-advocacy: it's clear that there is extensive resistance to stereotypes emerging among groups of elders active in public life, especially since 2000. For example, the head of an elder activist group states that "elders ... have the common sense, wisdom, resources (networks, skills, passion) and the inner urge to leave a legacy that [they] can be proud of" (Isner, 2015). This sense of concern for the future, and of a larger responsibility to next generation(s), has been well articulated by Phil Kingston, the 79-year-old cofounder of the organization Grandparents for a Safe Earth, and he has expressed concern that coming generations will likely judge us unfavorably. His logic for engaging with environmental activism later in life is that "we [elders] have the freedom most don't have: and let's use it, because there is not much they can do to us. We have very little to be afraid of" (Wayman, 2015). Kingston continued, lamenting the fact that after being arrested in a Barclay's bank following sit-in protests over the bank's support of fossil fuel extraction industries, he was promptly "de-arrested" on the street in front of the bank. Isner and Kingston are two of a growing number of retired American baby boomers, a group, according to a 2010 study, that grows by roughly 10,000 every day (Heimlich, 2010). According to the same study, baby boomers (then) also accounted for 26 percent of the total U.S. population. Another study predicts "with the baby boom in retirement by the year 2030, one in every five U.S. residents will be over 65" (Walls, 2014). For those reasons, climate change activist Bill McKibben specifically identified the boomer generation as a critical component of the fight to head off climate change (McKibben, 2008).

Seniors already contribute heavily to advocacy and social justice efforts. Some things that help drive them may include their lived experience (during which many have witnessed climate change effects), a sense of urgency (because of a shorter time horizon), freedom from worries about job security, flexible time and financial security. Many seniors came of age, politically and otherwise, around the time of the civil rights movement (Roszak, 2009); this may mediate in favor of more attunement to the core focus of the environmental justice movement. Given these attributes, seniors may have much to offer for advocacy, especially environmental advocacy, and there are many groups of seniors targeting just that. For example, the organization Gray Is Green describes itself as "an online gathering of older adult Americans aspiring to create a green legacy for the future" with a mission to "co-create a future of economic justice, ecological sustainability and social justice" (Gray Is Green, 2016). Another example is the organization Elders Climate Action, whose members describe themselves as the "guardians of future generations" with a goal of mobilizing current leaders and policy-makers to act more sustainably (Elders Climate Action, 2017). Elders Climate Action is a subsidiary project of the Conscious Elders Network, a larger organization that recognizes elders' "responsibility to do better, to bring more awareness and consciousness to protecting the greater community of life and its future generations" (Conscious Elders Network, 2017).

Some researchers have argued that for elder women specifically, activism is positive for mental health and empowerment, presenting data to support their assertions (McHugh, 2012).

The large number of senior-oriented organizations that are working toward mitigating the harms of climate change contravenes any notions of seniors as self-interested individuals, shamelessly consuming finite resources that could otherwise be preserved for the future; a more accurate picture might be that they are on the front lines trying to protect them. The collective message of many of these elder organizations and networks seems to be consistent: take accountability for damage being done to the Earth using collective wisdom and passion to leave behind an environmentally sustainable legacy that future generations can appreciate and cultivate.

Another aspect of elder identity is a person who has a life's work of accomplishments, possibly decades long, which he or she draws from in later projects. In that spirit, John Hart (2006) wrote *Legacy: Portraits of 50 Bay Area Environmental Elders*, which profiles the work of dozens of accomplished individuals who did environmental advocacy later in life. As Hart points out, the work of early advocates is underappreciated, and this extends to the Bay Area's natural beauty. He stresses: "This legacy thrives, in most cases, not because benevolent governments have provided for it, but because certain stubborn citizens felt called upon to work for its preservation, often against great odds," (Hart, 2006, Foreword).

Federal agencies have also jumped into the elder–environment nexus, with projects that seek to "help" elders as well as those that try to stimulate self-advocacy. Examples of such "helping" projects would be those that seek to improve health status and engagement in elderly populations. For example, the CDC's Healthy States Initiative (CDC, 2007) advocates for state and local entities "to provide volunteer opportunities for older adults … [as] volunteerism is a proven component of good health and quality of life" (CDC, 2007). The CDC has also developed a guide for carrying out community-based physical activity programs for seniors, which provides a five-element development process for creating effective programs called "RE-AIM" (Belza & PRC-HAN, 2007). The framework suggests specifying, "reach, effectiveness, adoption, implementation and maintenance" for new projects. The U.S. Department of Health's Administration for Community Living has a civic engagement initiative meant to attract volunteers to help run some of its aging programs, but also to "provide expanded opportunities for older adults and others to get involved in their communities" (Administration for Community Living, 2016).

An example of the second type of program, stimulating networking and advocacy, would be the EPA's listserv on aging and sustainability. This listserv promotes information exchange as well as opportunities for "environmental stewardship that contribute to a more sustainable world" (Sykes and Pillemer, 2014). More specifically, the EPA offers positions to retired or unemployed Americans (over age 55) in environmental fields through the

Senior Environmental Employment (SEE) program, engaging seniors with technical or professional skills in an environmental field, or those with clerical and non-clerical skills to assist in some capacity. Although the positions do not constitute federal employment, participants are compensated for their time (USEPA, 2016).

Considerations among distinct populations of elders

African-American communities have long been subjected to many types of discrimination and assault, targeting both the people and the places they live (the built environment). Elder African-Americans have been most directly and personally connected to these processes and transformations, especially since many lived through the era of segregation. African-American communities have been built, (re) built and sometimes even destroyed in the context of racist and racialized social practices (Kendi, 2014). Policies of the Jim Crow era shaped past practices and were written into institutions and, in turn, into the built environment itself. Many aspects of cities are taken as givens when in fact they resulted from political and administrative decisions reinforcing racial supremacy; while the predatory and exclusionary "redlining" practices affecting home sales are among the best known, others exist as well (Kendi 2014). By built environment, we mean the buildings themselves as well as the connections to one another and to the natural environments. Access to green amenities has been shaped by these practices as well (Finney, 2014).

These histories of African-Americans, green space and waterways suggest that the story centers on being shut out of mainstream public life; however, there's another dimension that needs to be mentioned. It would be a mistake to overlook the vitality and accomplishments of African-Americans in providing for themselves despite all the social and institutional factors working against them, and also in resisting authority asserting racist policies. It is worth acknowledging that in some places, in the pre-civil rights era, black-owned clubs thrived, providing beach access and nightclub entertainment in locations including Maryland's Chesapeake Bay area and the Mississippi River waterfront (Gadsden 2014; Kahrl, 2012). Historian Andrew Kahrl traced the foundations of "coastal capitalism" in which many properties once owned by African-Americans were subjected to dispossession, relegating local people to worse areas (more dangerous and polluted) for waterfront recreation (Kahrl, 2012).

In many communities, the civil rights-era battles for access to the public sphere included a chapter at a beach or swimming pool for basic recreation, and thus those sites became places where old racist stereotypes and fears were subjected to closer scrutiny, struggled over and sometimes transcended (Hastings et al., 2006). It is elders in the black community who know these stories personally and may share them with friends and family. The academic literature on these topics is not extensive and yet these histories are an important part of understanding what constitutes ethics and how struggles for justice take place. Stories of strength, ingenuity and resistance are important to access while it remains possible to learn them directly.

Some projects actively cultivate the connections between senior African-American residents, their communities and their memories. In Portland, Oregon, the SHARP study—Sharing History Through Active Reminiscence and Photo-imaging—is taking volunteer participants on walks through their neighborhoods to discuss and reminisce about the past they remember. They move in small groups through historically black areas previously filled with black-owned homes and businesses but now changed through gentrification. The project leader, Dr. Raina Coff, said, "A lot of our wisdom and stories about what community means comes from our elders," (Neergaard, 2017). The other goal of the project is to study whether "jogging memories where they were made can help older African-Americans stay mentally sharp and slow early memory loss" (Neergard, 2017).

In North America, Native American communities typically draw upon traditions in self-governance, and elders play a pivotal role in guiding their decision-making (Dunbar-Ortiz, 2014). In many indigenous societies, the concept of elder is specific and distinct, as a carrier of culture, knowledge and tradition; however, not all older people are elders, as it consists of both mastery and community recognition. This is especially important in indigenous communities since their own continuity depends on this transmission of culture. The persistence and survival of indigenous communities despite the genocidal action of governments is a testament to resistance. Educators seeking to teach about indigenous knowledge (especially environmental) have sometimes invited elders to come tell stories to students in classrooms; authors have also noted that both the knowledge and its transmission are quite complex and sophisticated, more so than it might seem at first (Ferreira et al., 2014). The observation that both the content and the way it is transmitted matter helps underscore how traditional ecological knowledge is deeply embedded within indigenous culture, and how efforts to pull out parts of it for a purposeful goal (such as stream restoration) would engender difficulties. To designate their understanding about the world as a "knowledge system" might inadvertently detract from the spiritual and cosmological character of their perspective; perhaps the term "belief system" is more accurate (Ermine, 1995).

Senior participation in environmental programs

Senior citizens have contributed to ongoing stream restoration projects in a variety of roles; however, their impact has been best documented in the area of stream monitoring. Many senior care centers partner with educational institutions to provide training or certification for particular environmental initiatives and a framework for volunteer work to be conducted, which may then be carried out under the supervision of the certifying institution or in smaller, localized groups. The contributions of these initiatives to environmental stewardship have been tremendous; we discuss several of these initiatives below.

One prominent example is the Pennsylvania Senior Environment Corps (PASEC), whose members volunteered a combined 15,000 service hours in its first year of operation alone (1997–1998) (*PR Newswire*, 1998). According to the same article, Tom Ridge (then-governor of Pennsylvania) endorsed the group with the following written statement:

> Pennsylvania has become a leader among states in changing how government works and thinks about protecting our environment. The dedicated and thoughtful actions by older Pennsylvanians—who have taken up the banner of protecting water resources—have helped make us a leader. The Pennsylvania Senior Environment Corps is proving to be a national model with unlimited potential to help communities address pressing local environmental problems.
>
> (*PR Newswire*, quoting Governor Tom Ridge, 1998)

According to Pennsylvania's 2003–2004 "Green Plan," PASEC's 2,000 active volunteers participated in 12,250 stream, groundwater and watershed monitoring events during the year 2003 (Governor's Green Government Council (GGGC), 2004). Since the organization's formation in 1998, volunteers have contributed more than 2 million service hours in total, at a value to the state of roughly $3 million per year (Nature Abounds, 2017).

The Environmental Alliance for Senior Involvement (EASI) claims to be the largest senior environmental action network in the world, and provides the model for Pennsylvania's Senior Environment Corps. Its website also highlights the Oklahoma Senior Environment Corps (SEC), which partners with Oklahoma Blue Thumb to monitor water quality, and the Virginia SEC, which currently operates in eight of Virginia's 14 watersheds (EASI, n.d.). A number of programs are described on EASI's website, including an environmental monitoring program where "volunteers take samples of water from local streams and conduct streamside analyses of such components as acidity, dissolved oxygen and temperature [and] also observe the stream habitat and assess aquatic life found in the stream" (EASI, n.d.). Contributions to environmental restoration programs are specified as interpreting brownfield and Superfund site data to determine if public interests are being accommodated, as well as "plant[ing] trees as buffer strips along waterways or roadways to address water, noise or air pollution [and maintaining] watches on sensitive ecosystems, report[ing] potential pollution problems to the appropriate authorities" (EASI, n.d.). The website also describes various statewide programs' contributions to water quality monitoring and watershed restoration efforts, along with a month-long water monitoring project that concludes on World Water Monitoring Day. This project was developed in partnership with a range of national and global organizations, including

America's Clean Water Foundation, the International Water Association and several federal agencies including USEPA, USDA and USGS.

The success of Pennsylvania's SEC has made this particular state-level SEC an ideal model for similar programs, and it has been used as an example for 35 state and 30 country SEC programs (GGGC, 2004). One organization patterning other statewide programs on the PASEC model is Nature Abounds, a national 501(c)(3) nonprofit. Nature Abounds created a partnership with the Chesapeake Bay Trust in Maryland to form an SEC that will assist with the ongoing revitalization project for the Chesapeake Bay, as well as planning expansion into New Jersey, Alabama, Montana and Florida (Nature Abounds, 2017).

Many seniors have made environmental initiatives their retirement projects. *The Wall Street Journal* published a story about a number of case studies that looked at seniors who had been successful in a range of fields and decided to devote their free time in retirement to various volunteer causes. One person profiled, Dr. Bea States, taught at the University of Pennsylvania for nearly three decades and, in retirement, turned her attention to the Peak Center's Senior Environment Corps. This 82-year-old volunteer holds a Ph.D. in biochemistry, and her projects there include testing and monitoring water supplies in northern Pennsylvania for environmental contaminants (Essick, 2009).

Master Naturalist Programs (MNP) all over the country target retirees as well as late-career professionals to engage in volunteer work geared toward environmental stewardship. MNPs are comprehensive groups that seek to "bridge the gap between conservation science and practice" (Merenlender et al., 2016). In a typically structured MNP, participants complete 40 to 50 hours of certification training over the course of a year and are expected to complete 40 hours of service within the next year. Partnerships among local universities, conservation organizations, land trusts and state/federal agencies are often utilized for funding purposes and to give volunteers ample exposure to technical skills during their training, which they will use for the duration of their volunteer involvement. Many of the MNP websites have bulletin boards and announcement pages where volunteers are notified of opportunities for additional trainings and courses that are offered through various groups. For example, Maryland's MNP describes a mission of "engag[ing] citizens as stewards of Maryland's natural ecosystems and resources through science-based education and volunteer service in their communities" (Maryland Master Naturalist Program, 2017). Virginia, California and Texas (to name a few) provide a similar structure for their MNPs, where the programs are overseen at the state level but managed and implemented at the local level. The Texas MNP currently has 48 "local" chapters of varying size; Region 7 accounts for the area within one county, whereas Region 36 is responsible for 26 counties (Texas Master Naturalist, 2017). The difference in area covered between these two chapters is significant. Maine's Naturalist Program keeps an updated list of naturalist training courses on its website, and specifies the location of five senior colleges and programs geared to seniors over 50 (Maine Master Naturalist Program, 2017).

The LIFE project, overseen by the University of Maryland, hopes to resolve current environmental challenges (particularly with regard to water quality) by engaging the aging population of baby boomers, which is currently "tilting the U.S. population toward older age," to address these challenges through training in environmental stewardship (Manning et al., 2010). The University of Maryland provides intensive training and support to participants throughout the course of their service. Those who complete the training go on to "take leadership responsibility for a wide array of projects designed to conserve, restore and enhance habitat in the Chesapeake Bay Watershed" (Manning et al., 2010). Antioch University in Seattle, Washington, offers a similar certification program, but in addition to targeting seniors and retirees, the Legacy Leadership program is intended for people who want to transform their workplace skills to the nonprofit sector (Taylor, 2007). Although its requirements are not as precise as SEC requirements, which specify that participants should be over the age of 55, this program seems oriented toward people seeking meaningful work in the second half of their lives (Antioch University, 2008).

Senior citizens seem to be seeking roles in environmental stewardship, including stream restoration and monitoring projects such as The Crabby Creek Initiative, an ongoing restoration project in a suburb of Philadelphia. Once they received some basic training, these community participants were highly effective in water quality monitoring roles, and their environmental attitudes changed as a result of their participation. Stream monitoring is a critical contribution to stream restoration projects as many grants for restoration efforts will fund the initial cleanup and revitalization of the waterway but may not provide funds to measure the effectiveness of the effort and ensure that water quality is maintained into the future. Such projects may rely heavily in the future on volunteer citizens to monitor the continued viability of these projects long after active restoration efforts have ended. However, diversity among the volunteer population in this organization was not high: a survey of the demographics of the participants in ongoing water monitoring for the project indicated that the majority of community volunteers involved with the project were financially stable (middle to upper class), white (Caucasian), female and over the age of 50 (McHugh, 2012).

Conclusions

When it comes to participating in the restoration of waterways, older is not necessarily "lesser." Furthermore, there is a good convergence between the life experience of many of today's elders and the skills needed for participation in restoration projects. Since many elders might have familiarity with "movement" politics, especially social movements focused on the environment or civil rights, they are well positioned for action in restoration projects requiring an advocacy orientation. The role of memory cannot be overstated in restoration work, and people with direct experience of a place and personal connections to the waterways can provide unique information, all of which is essential for restoring community connections to waterways (Krasny & Tidball, 2012). Often these elders hold knowledge of the past to the point that they actually *are* the record of what has been: they may possess precious information about material practices that doesn't exist anywhere else, and that may be accessible only in the context of projects focused on oral history, etc. The special identity of elders in certain communities (for example, Native American, African-American, immigrant) means that there may be distinct roles for them relative to younger adults.

References

Achenbaum, W.A. (1978). *Old Age in the New Land*. Baltimore, MD: Johns Hopkins University Press.

Achenbaum, W.A. (2005). *Older Americans, Vital Communities*. Baltimore, MD: Johns Hopkins University Press.

Achenbaum, W.A. (2010). Past as prologue: Toward a global history of ageing. In D. Dannefer & C. Phillipson (Eds.), *The SAGE Handbook of Social Gerontology* (pp. 21–32). London: SAGE Publications Ltd.

Administration for Community Living. (2016). Volunteer opportunities and civic engagement. U.S. Department of Health and Human Services [Website]. Retrieved from www.acl.gov/programs/volunteer-opportunities-and-civic-engagement

Beierle, T.C., & Konisky, D.M. (2000). Values, conflict, and trust in participatory environmental planning. *Journal of Policy Analysis and Management, 19*(4), 587–602. doi: 10.1002/1520-6688(200023)19:4<587::AID-PAM4>3.0.CO;2-Q

Belza, B. & PRC-HAN Physical Activity Conference Planning Workgroup. (2007). *Moving ahead: Strategies and tools to plan, conduct, and maintain effective community-based physical activity programs for older adults*. Atlanta, GA: Centers for Disease Control and Prevention. Retrieved from www.cdc.gov/aging/pdf/community-based_physical_activity_programs_for_older_adults.pdf

Brody, S.D., Highfield, W., & Peck, B.M. (2005). Exploring the mosaic of perceptions for water quality across watersheds in San Antonio, Texas. *Landscape and Urban Planning, 73*, 200–214. doi: 10.1016/j.landurbplan.2004.11.010

Bronx River Alliance (2017). History. Retrieved from http://bronxriver.org/?pg=content&p=aboutus&m1=44

Byron, J. (2004). Transforming the Southern Bronx River Watershed. Paper presented at the Walk 21-V Cities for People Conference [June 9–11, 2004, Copenhagen, Denmark]. Brooklyn, NY: Pratt Institute Center for Community and Environmental Development.

Campbell, L.K. (2006). Civil society strategies on urban waterways: Stewardship, contention and coalition building. (Unpublished master's thesis). MIT, Cambridge, MA.

Centers for Disease Control (CDC). (2007). Keeping the *Aging Population Healthy*: Legislator *Policy Brief*. Retrieved from www.giaging.org/documents/CDC_Healthy_States_Initiative.pdf

Conscious Elders Network. (2017). Mission. Retrieved from www.consciouselders. org/mission

de Beauvoir, S. (1972). *The Coming of Age*. New York, NY: G.P. Putnam and Sons.

DeVillo, S.P. (2015). *The Bronx River in History and Folklore*. Charleston, SC: The History Press.

Dunbar-Ortiz, R. (2014). *An Indigenous Peoples' History of the United States*. Boston, MA: Beacon Press.

EASI. (n.d.). Home page. Retrieved from www.easi.org

Elders Climate Action. (2017). Who We Are. Retrieved from www.eldersclimateaction. org/who_we_are

Ermine, W. (1995). Aboriginal epistemology. In M. Battiste & J. Barman (Eds.) *First Nations Education in Canada: The Circle Unfolds* (pp. 101–112). Vancouver: UBC Press.

Essick, K. (2009, October 17). Encore (A special report): Profiles in later life. *Wall Street Journal*. Retrieved from www.wsj.com/articles/SB10001424052748703298 004574457472455337390

Ferreira, M.P., McKenna, B., & Gendron, F. (2014). Traditional elders in post-secondary STEM education. *International Journal of Health, Wellness, and Society*, 3. Retrieved from http://digitalcommons.wayne.edu/nfsfrp/10

Finney, C. (2014). *Black Faces White Spaces: Reimagining the Relationship of African Americans to the Great Outdoors*. Lexington: The University of Kentucky Press.

Frumkin, H., Fried, L. & Moody, R. (2012). Aging, climate change, and legacy thinking. *American Journal of Public Health*, 102 (8), 1434–1438. doi: 10.2105/ AJPH.2012.300663

Gadsden, B. (2014, March 11). Race, capitalism, and the rise and fall of black beaches. Southern Spaces: A Journal About Real and Imagined Spaces and Places of the US South and their Global Connections. Retrieved from https://southernspaces.org/ 2014/race-capitalism-and-rise-and-fall-black-beach-communities

Gray Is Green. (2016). Who we are. Retrieved from http://grayisgreen.org/our-story/

Hart, J. (2006). *Legacy: Portraits of 50 Bay Area Environmental Elders*. San Francisco, CA: Sierra Club Books.

Hastings, D.W., Zahran, S. & Cable, S. (2006). Drowning in inequalities: Swimming and social justice. *Journal of Black Studies*, 36(6), 894–917. Retrieved from www. jstor.org/stable/40034351

Heimlich, R. (2010). Baby Boomers Retire. Pew Research Center. Retrieved from www.pewresearch.org/fact-tank/2010/12/29/baby-boomers-retire/

Hillman, M. (2004). The importance of environmental justice in stream rehabilitation. *Ethics, Place and Environment*, 7(1–2), 19–43. doi: 1080/1366879042000264750

Hillman, M. (2005). Justice in river management: Community perceptions from the Hunter Valley, New South Wales, Australia. *Geographical Research*, 43(2), 152–161. doi: 10.1111/j.1745-5871.2005.00310.x

Hogan, H., Perez, D. & Bell, W.R. (2008) Who (really) are the first baby boomers? *Joint Statistical Meetings Proceedings: Social Statistics Section*, Alexandria, VA: American Statistical Association, 1009–1016.

Hopkins, A.W. (2005). *Groundswell: Stories of Saving Places, Finding Community*. San Francisco, CA: The Trust for Public Land.

Isner, L. (2015). Becoming Elders in our Evolving World. Pachamama Alliance. Retrieved from www.pachamama.org/blog/becoming-elders-in-our-evolving-world

Kahrl, A. (2012). *The Land Was Ours: African American Beaches from Jim Crow to the Sunbelt South*. Cambridge, MA: Harvard University Press.

Kendi, I.X. (2014). *Stamped from the Beginning: The Definitive History of Racist Ideas in America.* New York, NY: Nation Books.

Krasny, M.E., & Tidball, K.G. (2012). Civic ecology: A pathway for Earth stewardship in cities. *Frontiers in Ecology, 10*(5), 267–273. doi: 10.1890/110230

Lynd, R.S., & Lynd, H.M. (1929). *Middletown.* New York, NY: Harcourt Brace.

McHugh, M. C. (2012). Aging, agency, and activism: Older women as social change agents. *Women & Therapy, 5*(3–4), 279–295. doi: 10.1080/02703149.2012.684544

McKibben, B. (2008). *The Bill McKibben Reader: Pieces from an Active Life.* New York, NY: St. Martin's Press.

Maine Master Naturalist Program (2017). About Us. Retrieved from http://mainemasternaturalist.org/course-overview/

Manning, T., Campbell, P. & Schmeckpeper, B. (2010). The Chesapeake Bay Legacy Institute for the Environment: Contributing to community, leaving something for the future. *Generations, 33*(4), 87–89. Retrieved from www.ingentaconnect.com/contentone/asag/gen/2009/00000033/00000004/art00016

Marshall, N. (2001). Bronx River Restoration: Report and Assessment. (Unpublished undergraduate thesis). Fordham University, Bronx, NY. Retrieved from http://fordham.bepress.com/environ_theses

Maryland Master Naturalist Program. (2017). Become a Master Naturalist. Retrieved from https://extension.umd.edu/masternaturalist

May, R. (2006). "Connectivity" in urban rivers: Conflict and convergence between ecology and design. *Technology in Society, 28*, 477–488. doi: 10.1016/j.techsoc.2006.09.004

Merenlender, A.M., Crall, A.W., Drill, S., Prysby, M. & Ballard, H. (2016). Evaluating environmental education, citizen science, and stewardship through naturalist programs. *Conservation Biology, 30*(6), 1255–1265. doi: 10.1111/cobi.12737

Metchnikoff, E. (1905). "Old age," *Smithsonian Annual Report, 1904–05.* Washington, D.C.: Government Printing Office.

Moran, S. (2007). Stream restoration projects: A critical analysis of urban greening. *Local Environment, 12*(2), 111–128. doi: 10.1080/13549830601133151

Moran, S. (2010). Cities, creeks, and erasure: Stream restoration and environmental justice. *Environmental Justice, 3*(2), 61–69. doi: 10.1089/env.2009.0036

Moran, S., Perreault, M., & Smardon, R. (2016). Finding our way: A case of waterway restoration and participatory process. *Landscape and Urban Planning,* doi: 10.1016/j.landurbplan.2016.08.004

Moyer, E. (2015, May 21). Elders Take Action on Climate Change. Huffington Post. Retrieved from www.huffingtonpost.com/ellen-moyer-phd/elders-take-action-on-climate-change-_b_7309578.html

Nassauer, J.I., Kosek, S.E., & Corry, R.C. (2001). Meeting public expectations with ecological innovation in riparian landscapes. *Journal of American Water Resources Association, 37*(6), 1439–1443. doi: 10.1111/j.1752-1688.2001.tb03650.x

Nature Abounds. (2017). Senior Environment Corps. Retrieved from www.natureabounds.org/SEC.html

Neergaard, L. (2017, July 25). Black seniors in Portland test reminiscing to guard against Alzheimer's. The Seattle Times. Retrieved from www.seattletimes.com/nation-world/nation-politics/black-seniors-stroll-down-memory-lane-aiming-to-stay-sharp/

Ortman, J.M., Velkoff, V.A. & Hogan, H. (2014). An aging nation: the older population in the United States—population estimates and projections. Retrieved from www.census.gov/prod/2014pubs/p25-1140.pdf

Pennsylvania Governor's Green Government Council [GGGC]. (2004). *Commonwealth of Pennsylvania: Green Plan 2003–2004.* Retrieved from www.elibrary.dep.state.pa.us/dsweb/Get/Document-46265/01%20GGGC-BK-DEP3102.pdf

Perini, K. & Sabbion, P. (2017). *Urban Sustainability and River Restoration: Green and Blue Infrastructure.* Chichester, U.K.: Wiley Blackwell.

Perreault, T., Wraight, S. & Perreault, M. (2012). Environmental injustice in the Onondaga Lake waterscape, New York State, USA. *Water Alternatives, 5*(2), 485–506.

Petts, J. (2006). Managing public engagement to optimize learning: Reflections from urban river restoration. *Human Ecology Review, 13*(2), 172–181.

Petts, J. (2007). Learning about learning: Lessons from public engagement and deliberation in urban river restoration. *The Geographical Journal, 173*(4), 300–311. doi: 10.1111/j.1475-4959.2007.00254.x

Pillemer, K., Wells, N.W., Meador, R.H., Schultz, L., Henderson Jr., C.R. & Tillema Cope, M. (2017). Engaging older adults in environmental volunteerism: The retirees in service to the environment program. *The Gerontologist, 57*(2), 367–375. doi: 10.1093.geront.gnv693

PR Newswire Association (1998, November 10). Pennsylvania senior environment corps has successful first year. Retrieved from www.thefreelibrary.com/Pennsylvania+Senior+Environment+Corps+Has+Sucessful+First+Year.-a053198140

Quadagno, J. (1989). Generational equity and the politics of the welfare state. *International Journal of Health Services, 20*(4), 631–649. doi: 10.1177/003232928901700303

Rogge, M.E., Davis, K., Maddox, D. & Jackson, M. (2005). Leveraging environmental, social, and economic justice at Chattanooga Creek: A case study. *Journal of Community Practice, 13*(3), 33–53. doi: 10.1300/J125v13n03_03

Roszak, T. (2009). *Making of an Elder Culture: Reflections on the Future of America's Most Audacious Generation.* Gabriola, British Columbia: New Society Publishers.

Runfola, A., & Weiss, J. (2007). *The Bronx River Classroom: The Inside Track for Educators.* New York, NY: The Bronx River Alliance and NYC Parks.

Schaie, K.W., & W.A. Achenbaum (Eds.). (1993). *Social Import on Aging: Historical Perspectives.* New York, NY: Springer.

Schauman, S., & Salisbury, S. (1998). Restoring nature in the city: Puget Sound experiences. *Landscape and Urban Planning, 42,* 287–295.doi: 10.1016/S0169-2046(98)00093-0

Selman, P., Carter, C., Lawrence, A., & Morgan, C. (2010). Re-connecting with a neglected river through imaginative engagement. *Ecology and Society, 15*(3). Retrieved from www.ecologyandsociety.org/vol15/iss3/art18/

Spirn, A.W. (2005). Restoring Mill Creek: Landscape literacy, environmental justice and city planning and design. *Landscape Research, 30*(3), 395–413. doi: 10.1080/01426390500171193

Sykes, K. & Pillemer, K. (2014). The intersection of aging and the environment. *Generations: Journal of the American Society on Aging.* Retrieved from www.asaging.org/blog/intersection-aging-and-environment

Taylor, K. (2007, July 27). Nature getaway is close as the Bronx. The Sun. Retrieved from www.nysun.com/new-york/nature-getaway-is-close-as-the-bronx/59305/

Taylor, L. (2007, October 22). Why retire just when the workplace is getting interesting? *The Seattle Times.* Retrieved from www.seattletimes.com/seattle-news/health/why-retire-just-when-the-workplace-is-getting-interesting/

Texas Master Naturalist (2017). Our chapters. Retrieved from http://txmn.org/chapters/

Tunstall, S.M., Tapsell, S.M., & Eden, S. (1999). How stable are public responses to changing local environments? A "before" and "after" case study of river restoration. *Journal of Environmental Planning and Management 42*(4), 527–547. doi: 10.1080/09640569911046

Tunstall, S.M., Tapsell, S.M., & Eden, S. (2000). River restoration: Public attitudes and expectations. *Water and Environment Journal, 14,* 363–370. doi: 10.1111/j.1747–6593.2000.tb00274.x

U.S. EPA. (2016). Senior Environmental Employment (SEE) Program. Retrieved from www.epa.gov/careers/senior-environmental-employment-see-program

Wade, R.J., Rhoads, B.L., Rodríguez, J., Daniels, M., Wilson, D., … Schwartz, J. (2002). Integrating science and technology to support stream naturalization near Chicago, Illinois. *Journal of the American Water Resources Association, 38*(4), 931–944. doi: 10.1111/j.1752-1688.2002.tb05535.x

Walls, D. (2014). Elders Rights Movement. Retrieved from www.sonoma.edu/users/w/wallsd/pdf/Elders-Rights-Movement.pdf

Wayman, S. (2015, July 14). Why our elders are becoming environmental activists. Irish Times. Retrieved from www.irishtimes.com/life-and-style/health-family/parenting/why-our-elders-are-becoming-environmental-activists-1.2275470

Wessels, A.T. (2010). Place-based conservation and urban waterways: Watershed activism in the bottom of the basin. *Natural Resources Journal, 50,* 539–560.

Wolch, J. (2007). Green urban worlds. *Annals of the Association of American Geographers, 97*(2), 373–384. doi: 10.1111/j.1467-8306.2007.00543.x

7 Community engagement and mapping

Richard Smardon

Introduction

The approaches to community engagement around stream restoration projects have evolved over time. The first approach was information dissemination—one-way communication that sought to inform. A latter stage has acknowledged more involvement, and a final stage has included decision-making and continuing two-way interaction. Participation is essential, as seen in the South Bronx and Onondaga Creek, Syracuse, NY. Still more advanced approaches acknowledge the differential types of collaboration: Some involve deciding together (shared decision-making). Others involve shared exploration for and discovery of facts and information, and these co-creation models are especially promising for environmental justice (EJ) goals. Another point that is especially important is mapping/ GIS; because of the quintessentially spatial nature of revitalization projects, having a graphic representation of connectivity is more productive. Several ways exist to accomplish this, including low tech, with big maps and stickers, or digitally supported/GIS/web-based. Finally, the use of smartphones and social media are the most dynamic strategies. These approaches are covered within this chapter.

The previous chapters create the foundation for more utilization of interactive participatory methods to engage urban waterway neighborhoods in collaborative and social learning processes as part of revitalization waterway planning (Petts, 2006, 2007; Smardon et al., 1996). Community engagement and mapping are both needed to incorporate risk from water quality threats and flooding, perceived value(s), perceptions of existing waterway qualities and understanding of hydrologic and ecological processes.

There is much published work on collaborative processes at the watershed or catchment scale for water resource planning in North America and Europe (Bos & Brown, 2015; Leach et al., 2002; Sabatier et al., 2005), but very little of it addresses the stream reach scale. Herringshaw et al. (2010) articulated the challenges for urban water quality planning as "a public that often lacks understanding of ecological principles, inadequate evidence of the effectiveness of restoration practices, and difficulty integrating social and biophysical factors" (Herringshaw et al., 2010, p. 535). So the issue here is that the professional or the scientist thinks the public just needs to understand the physical science aspects of stream restoration and revitalization without also understanding the local perception of these issues. Such approaches are doomed to failure in many cases.

Herringshaw et al. (2010) explained that a collaborative learning process over an extended period of time applied to implementation of an urban riparian buffer along a headwater stream in a neighborhood park increased participants' knowledge about water quality problems associated with urbanization, stormwater and nonpoint source pollution, as well as their perception of the importance of stream ecosystem functions. Riley (2016) pointed out that urban residents might place more value on safety and aesthetic concerns than ecological or hydrologic waterway restoration. Still, for practitioners, questions remain about which types of collaborative methods facilitate the best learning outcomes. Again professionals need to understand local resident perceptions, especially if there is a perceived or real environmental justice issue, which could be procedural or substantive.

The following sections will briefly outline the types of special data and engagement processes that have been applied to urban waterway revitalization followed by case study applications.

Review of community spatial data needs

Community spatial data is needed to understand historical environmental justice issues as well as current environmental and social conditions that have not been alleviated. Some of this spatial data is physical and some is social and psychological. There are at least three types of spatial data, which could include:

- *Physical data* such as water quality, flooding risk and biodiversity, as they affect human health and access. We have good information for why we should restore, revitalize or naturalize urban waterways (Findlay & Taylor, 2006), visualize human impacts on such systems (Kondolf et al., 2006) and develop criteria to evaluate restoration success (Woolsey et al., 2007). One example is hydrological data to show how more natural flows could be accommodated in relation to flooding issues in Wildcat Creek in Richmond, California.
- *Social data* as it relates to human health, access and environmental justice. We are just beginning to understand the connections between urban waterways and green infrastructure such as urban parks to human health (Barnhill & Smardon, 2012; Tzoulas et al., 2007). This is illustrated by gathering economic levels and racial data for areas subject to flooding and pollution, such as in the Anacostia River communities.
- *Psychological perception* as it relates to place-based values, understanding, needs and expectations. Rhoads et al. (1999) stated that

place-based contexts are important if there is to be positive interaction between scientists and non-scientists for community-based watershed management. It has been argued repeatedly that aesthetic improvement and recreational access are major variables affecting waterway residents' perceptions of environmental quality (Asakawa et al., 2004; Nassauer et al., 2001; Özgüner et al., 2012). One example here is the perception of Onondaga Creek in Syracuse, NY, as a dangerous creek during flood flow events by adjacent Southside residents.

Review of spatial engagement processes

For the purpose of engaging landowners and other stakeholders to understand revitalization or restoration processes, some standard techniques include demonstration projects; field days and tours; newsletters; displays in public places and events; talks to community groups; newspaper, radio and television stories; websites; social media; email; and work with schools and community groups (Souder, 2013). However, most of these techniques are not really interactive since they are predominantly one-way forms of communication.

The U.S. Environmental Protection Agency (USEPA) has also published a catalog of community engagement and research methods (USEPA, 2002). Meanwhile, the Agency has funded a decision support system for the selection and placement of best management practices (BMPs) for urban watershed improvement (Riverson et al., 2004). The objective of such a system is to provide stormwater managers with sound tools for developing, evaluating, selecting and placing BMP options, based on cost and effectiveness. However, no part of the decision support system includes public input as a pivotal element in the process. This is problematic in that USEPA guidance does not specifically address procedural environmental justice and representation of local issues. Meanwhile, there are examples of projects that work differently. In Philadelphia, planners used the strategy of "mediated modeling" to involve neighborhood residents in developing green infrastructure to alleviate stormwater impacts (Montalto et al., 2013; Zidar et al., 2017). Such an approach involves identifying local needs and issues, developing alternatives and getting reactions to these alternatives in terms of remediating EJ issues, cost and effectiveness.

A number of studies and projects in North America, Australia, Asia and Europe use geographic information systems (GIS) for watershed and larger waterway planning, but few of these GIS systems have been utilized at the waterway stream or creek level and very few in an urban context. Harrison et al. (n.d.) have developed a GIS system that can be used to estimate landscape indicators and target waterway restoration needs. This system has been utilized for targeting restoration needs for Brasstown Creek, South Carolina, sub-watersheds and describing riparian influences on sedimentation in the Chattanooga watershed in North Carolina. But these GIS mapping approaches have been predominantly applied to larger-scale water-shed and sub-watershed areas with biophysical data and not social and psy-chological data, which might yield the existence of local EJ issues.

Newer, more collaborative approaches include participatory GIS, agent-based modeling, participatory design and crowdsourcing, which will be briefly reviewed below. Environmental information systems and GIS have been utilized to build historical and future land use and natural resource conditions for the western Willamette River Basin in Oregon (Hulse et al., 2004). Guzy et al. (2008) have utilized agent-based modeling to assess future impact of urban expansion at the juncture of the McKenzie and Willamette rivers in Eugene and Springfield, Oregon. Both of these studies utilize public input in the development of future scenarios.

Agent-based modeling (ABM) allows physical and social/cultural environments to be modeled concurrently, and shows complex interactions among subsystems. Autonomous agents are recruited within an ABM to adhere to a set of behavioral rules. Such agents may learn based on changes within the system, or remain fixed. The agents react with one another as well as their environment so patterns emerge and the system evolves. ABMs are used as a predictive tool and to explore relationships among different agents and groups (Zidar et al., 2017). ABM was extensively used to model decisions about implementing green infrastructure in Point Breeze, Philadelphia (Montalto et al., 2013; Zidar et al., 2017). The critique of agent-based modeling is that it may not truly capture local grassroots issues, especially EJ issues.

Ghaemi et al. (2009) have developed a web-based platform to support interactive environmental planning in Southern California. The Green Visions project objectives include promoting more equitable park and open space access for local residents, protecting and restoring biodiversity and protecting and enhancing watershed health. The Interactive Park Analysis tool operates at the parcel level scale and provides access to large detailed geospatial data sets.

GIS and decision-support systems are moving more and more toward public accessibility. A new subfield of GIS is Participatory GIS or PGIS, which allows individuals to supply geo-specific data to the geographic system and/or have public access to the system. As an example, Cutts (2013) described a study that utilized a participatory GIS to map public outreach and com-pare differences in information availability across metropolitan Phoenix, Arizona. Dernoga et al. (2015) described developing using environmental justice indicators and assessment methods with GIS software to layer map-relevant EJ and watershed health indicators for Baltimore County, Maryland. Data included demographic characteristics, human health indicators and watershed health indicators. Highlighted results included several watersheds where communities were at risk for environmental justice issues associated with water quality, which would be either combined sewer outflows or breaks in the system within low-income neighborhoods.

Community participatory mapping can also be utilized for monitoring environmental quality. In this application one would expect that residents would be willing to provide geo-specific data about environmental quality such as spills or other pollution events. Im (2011) described such a community-mapping project utilizing smartphone technology with GPS to upload environmental information that then can be viewed on an interactive website or smartphone. Axler et al. (2006) described a Lake Superior watershed web-based real-time stream monitoring system. The website delivers real-time data values for flow, temperature, turbidity and conductivity in conventional formats and via a data animation tool from sensors in three urban trout streams including the St. Louis River discharge to Lake Superior and two North Shore tributaries.

Moving to social media, the question is whether social media like Facebook and other such platforms could be used to reach out to communities to collect data, discover local issues and promote dialogue concerning local waterway revitalization. Rivera et al. (2014) proposed a crowd-based system for Green Infrastructure (GI) design. They argued that such systems are in need of a participatory framework that considers community-specific social, cultural, economic and political constraints. The National Science Foundation (NSF)-funded project seeks to develop such a system integrating interactive, neighborhood-scale collaborative design by multiple stakeholders. Research tasks for the NSF-funded project include:

- Creation of integrated models to predict hydrologic, human and ecosystem impacts from GI designs at site to catchment scale;
- Development of interactive methods for crowdsourcing model parameterization for GI design; and
- Implementation of modeling and crowdsourced design methods in a cyber infrastructure framework.

Below is one case study that illustrates how interactive frameworks have been used for collaborative waterway revitalization planning.

Case study: Onondaga Creek participatory mapping and project development

The following case study is a summary of the interactive mapping exercise that was conducted as part of the Onondaga Creek Conceptual Revitalization Plan (OEI, 2009). Parts of the text come directly from the final draft plan (OEI, 2009, pp. 36, 44). Other steps in the collaborative planning process are included in Moran et al. (2013). The Onondaga Creek Working Group was made up of citizens who represented different creek communities from the rural headwaters to the Onondaga Nation to the inner city of Syracuse, New York, and who worked together for the duration of the project from

2005 to 2008. This project inherited two major environmental justice issues (Perreault et al., 2012, p. 213):

- The issue of historical degradation of the Onondaga Creek watershed, which is sacred and part of the Onondaga Nation creation story; and
- Near Syracuse Southside, where creekside residents had been subject to flooding damage, combined sewer overflow (CSO) water pollution and lack of access to the creek.

To facilitate the Onondaga Creek Working Group's design charrette, the Onondaga Environmental Institute (OEI) created a set of planning maps, 8 to 10 feet long, from aerial images of the Onondaga Creek corridor and its tributaries. OEI also developed a set of 40 cards with graphic representations (symbols) of creek revitalization options (see Figures 7.1 and 7.2). The symbol cards were based on options discussed by the Working Group, gleaned from community input and from references on stream restoration practice (Center for Watershed Protection, 2004; Federal Interagency Stream Restoration Working Group (FISRWG), 1998; Kloss et al., 2006; Pinkham, 2000). In addition to the symbol cards, the Working Group used blank cards and markers to customize maps.

The Working Group worked on the maps over two meetings (see Figure 7.3). They split into three teams: urban, rural and "mixed" land use. The urban team placed ideas on maps of the creek corridor from the Inner Harbor to Ballantyne Avenue. The mixed or transitional team placed ideas on two planning maps: Ballantyne Avenue to the northern border of the Onondaga Nation, and the Furnace Brook corridor. The rural team covered the remaining segments. Three team facilitators with community design experience were invited to facilitate each team during map making. The resource experts who assisted with options development were invited to return and advise the teams. For the planning map representing the Onondaga Nation territory area, Jeanne Shenandoah facilitated input from members of the Onondaga Nation. Sticky notes were used instead of the symbol cards. This is critical, as the Onondaga Nation has suffered EJ issues of a flood control dam constructed on sovereign territory as well as degradation of the whole watershed; yet they adapted to this participatory graphic process in their own way.

Map review and project area development

The large planning maps were then converted into digital representations by OEI. Symbols, notes and additional drawings were reproduced on the digital versions as placed by the Working Group on the original planning maps. Working Group members each received a tabloid-size set of the planning maps to verify and review. The Project Team grouped revitalization map ideas into project areas. The bundles represented future potential project

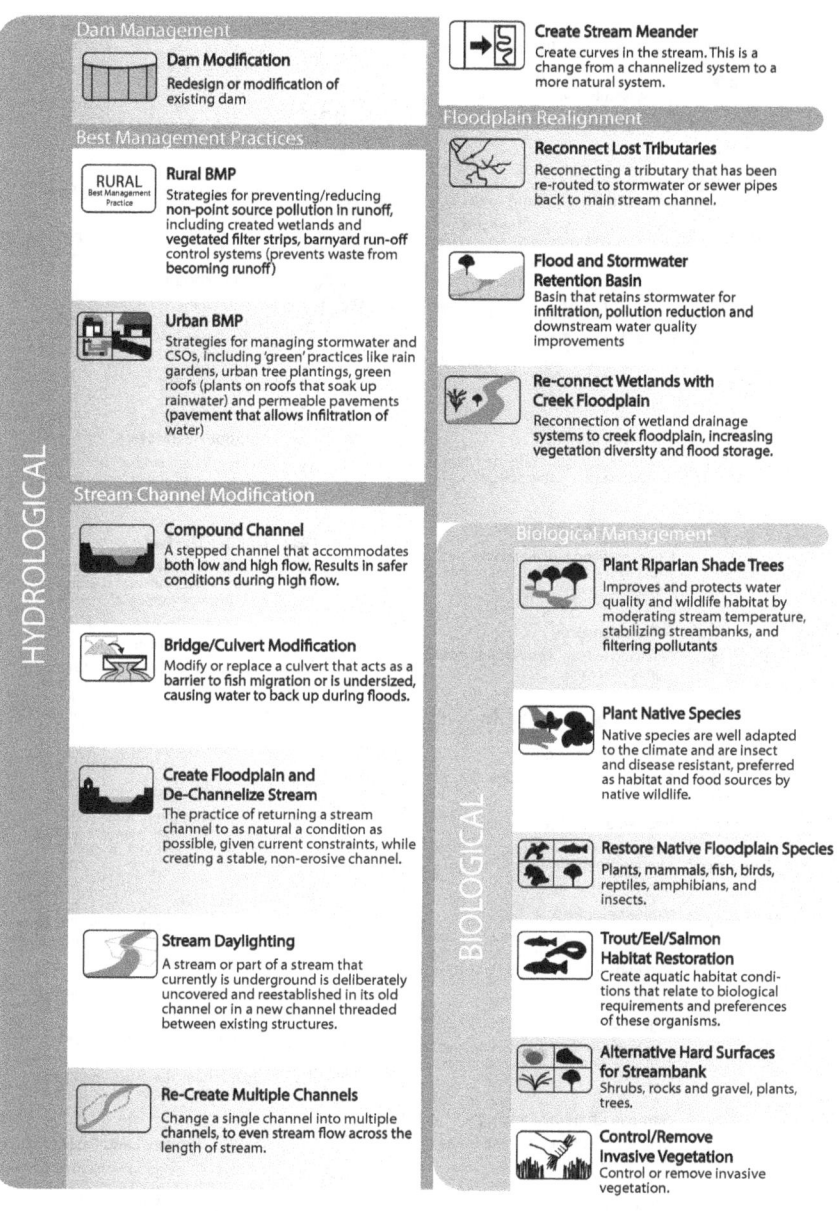

Figure 7.1 Hydrological and biological playing cards
Source: OEI (2009, p. 46), redrawn by Ryan Mackerer

Create/Manage/Restore Wetland
Either emergent wetland with grassy/shrubby vegetation, or forested wetland with tree species adapted to wetland soil types.

Create/Restore Upland Area
Higher areas upslope of streams, wetlands and riparian zones.

Remove Overgrowth Above/Around Creek
Remove excess invasive vegetation.

Kayak/Canoe/Boat Access Point

Paved or Gravel Foot/Bike Path

Land Management

Urban Ecopark
A community of businesses that enhance environmental and economic performance by collaborating to manage raw materials, energy, water, and waste.

Cultural/Historic Site
Buildings, sites, land of cultural or historical importance, open to visitation.

Multiple Use Park
Offer open space and recreational opportunities, includes visitor facilities and site improvements.

Urban Creek Preserve
Similar to a Bio Preserve but set in an urban environment. Undeveloped greenspace with minor improvements, facilities. It may be used to connect other greenspace and corridors.

Scenic Use Area
Natural vegetation, some social encounters, some visitor facilities, designed for outdoor recreation.

Bio Preserve
Natural vegetation, few social encounters, designed to preserve native plant and animal communities.

Land Acquisition

Creation of Public Park Land
Land purchased by a municipality or organization, managed and kept in a natural state, accessible to the public.

Purchase Private Land Easement
Includes conservation easements, a legal agreement between a landowner and an organization or government that prevents development or preserves scenic natural values of the land.

Safety/Flood Management

Natural Fence/Barrier
Shrubs, trees or vegetation planted next to the creek, as a barrier.

Improve Lighting
Increase lighting for safe use without causing harm to other species.

Flood-Proof Buildings
Flood-proofing individual structures with barriers, door dams and other measures.

Safety Measures
Can include high water warning lights, signage, fencing.

Recreation/Access

Fishing Access Point
Create public fishing access.

Pedestrian bridge
Bridge restricted to motor vehicles, intended for pedestrian/bike use.

Whitewater Park
Intended for kayak/canoe access, can include construction of stream features that enhance whitewater recreation.

Signage
Can include educational kiosks, nature trail, and directional types of signs.

Remove Chanlink Fencing
Create access or replace with more aesthetic options.

Nature Trail
A natural trail with small interpretive / educational signs.

Figure 7.2 Biological and land use playing cards
Source: OEI (2009, p. 47), redrawn by Ryan Mackerer

Figure 7.3 Workshop charrette using project cards and maps
Source: R. Smardon

areas for implementation of revitalization projects. OEI developed themes for each project area based on symbol groupings. The Working Group reviewed and voted on their preferred potential project areas bundled into interconnected projects. This was part of the participatory process related to decision-making and prioritization.

Revitalization map series results

The Working Group used symbols generated by OEI to represent ecological, hydrological, land use and green space actions during the design charrette process in May and June of 2007. The symbols were placed on a series of planning maps during the charrette as described previously. Project areas on the revitalization maps featured groupings of symbols suggesting specific areas of work. Adjacent to the maps were synopses of each project area based on the revitalization map results and notes taken during the design charrette. The project areas did not reflect land purchases, but rather represented areas of focus for future revitalization work.

The potential project areas were created for two reasons: First, grouping symbols into project areas ascribed them a recognizable identity for funding

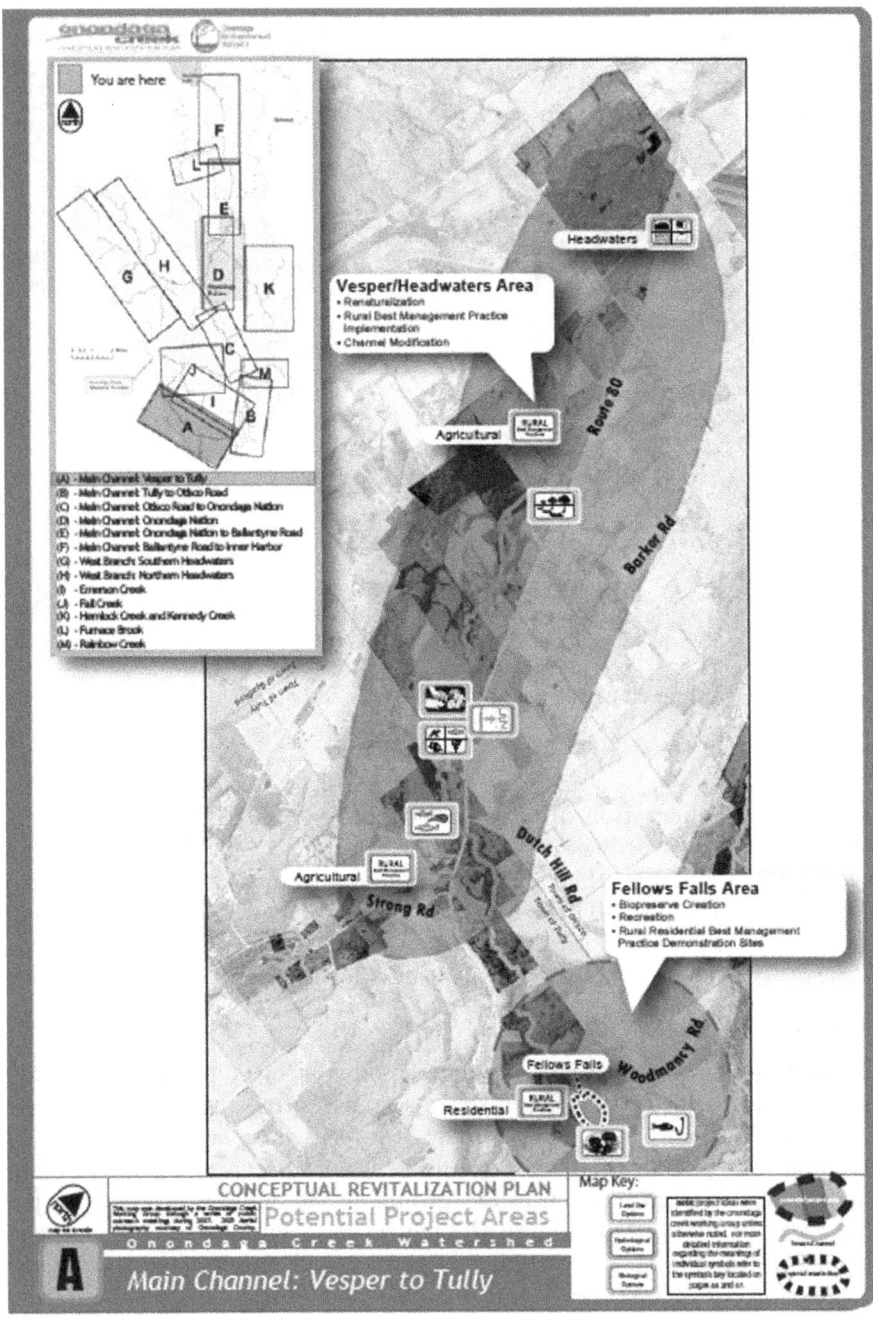

Figure 7.4 Sample final project map: Main Channel—Vesper to Tully
Source: OEI (2009, p. 49)

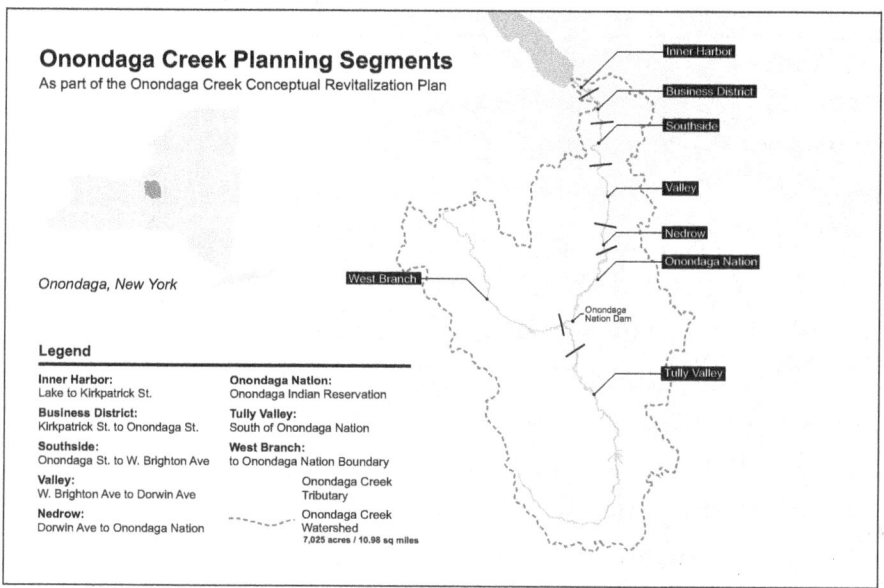

Figure 7.5 Onondaga Creek planning segments
Source: Drawn by Mark Warfel, Jr.

and building public support; second, project areas grouped revitalization ideas so that they were conceptualized holistically. Some recommendations would be easy to implement and others more difficult. Grouping easy and difficult ideas together served as a reminder that the creek and its surrounding watershed are an ecosystem that should receive full revitalization, rather than simply completing "cosmetic" treatments, leaving more difficult projects undone.

The revitalization maps are conceptual (see Figures 7.4 and 7.5). Just as at the public input meetings, the Working Group was asked to "think big" about the future of Onondaga Creek. Ideas on paper help move the community and decision-makers to revitalization actions. As stated in the Working Group's goals, a balance between use and protection has to be achieved. Community input indicated that forms of recreation, followed by a clean creek with natural areas and fishing opportunities, were the most frequent goal themes for Onondaga Creek. Striking the balance between use and protection will require accommodating the following factors: increasing recreational opportunities, ensuring clean water, protecting natural areas in the watershed and respecting the rights of private landowners.

Attendees of the October 2007 Working Group meeting placed sets of stars directly on the original revitalization maps to "vote" for their preferred potential project areas. Absent Working Group members were also given the opportunity to send in their votes by mail. Voting results (see Figure 7.6)

Potential Project Area	Map	Votes
Urban Section		
Southside Area	F	21
Botanical Garden Area	F	20
Inner Harbor	F	18
Armory Square	F	12
Clinton Square	F	11
Franklin Square	F	8
Furnace Brook Daylighting Project	F	7
Transition Sections		
South Valley Area	E	31
North Valley Area	E	25
Valley Watershed Biopreserve	E	20
Furnace Brook Watershed	L	20
Rural Sections		
Onondaga Nation Area	D	14
Honeywell Lands South	B,I	13
Fall Creek Area (Blue Hole)	J	12
Mudboils Area	B	8
LaFayette Apple Festival	C	7
Rainbow Creek Area	M	7
Vesper/Headwaters Area	A	6
Kennedy Creek Area	K	6
South Onondaga Area (W. Branch)	G,H	6
Fellows Falls Area	A	5
Honeywell Lands North	C	5
Central LaFayette Area	K	5
Pumpkin Hallow Area (W. Branch)	G,H	5
Tully Farms Byway Signage Project	C	4
Headwaters Gravel Mine	B	2

Figure 7.6 Potential projects and vote tally
Source: OEI (2009, p. 45)

reflected the Working Group's determination for which project areas best reflected plan goals and priorities. In the urban, transition and rural sections, the Working Group preferred the Southside Area (an EJ area), the South Valley Area and the Onondaga Nation Area (an EJ area), respectively. Created images, or renderings, that represent ideas from the revitalization maps were included within the final conceptual revitalization plan for the Southside Area and the South Valley Area (OEI, 2009).

Summary

If we are to successfully engage urban waterway neighborhoods in more meaningful revitalization or naturalization, more interactive processes are needed from the beginning through implementation. Such processes and interactive platforms must address place-specific environmental justice issues as well as potentially conflicting goals, e.g., safety line of sight versus aesthetics versus sustainable ecologic/hydrologic waterway design interventions. The good news is that community mapping and other interactive platforms are being developed, tested and implemented. The bad news is that such platforms and interactive processes may not be available for the most needy communities. Academics and communities can partner via service learning to address this issue of resource accessibility.

References

Asakawa, S., Yoshida, K., & Yabe, K. (2004). Perceptions of urban stream corridors within the greenway system of Sapporo, Japan. *Landscape and Urban Planning*, 68(2–3), 167–182. doi: 10.1016/S0169-2046(03)00158-0

Axler, R., Hagley, C., Host, G. & Schomberg. J. (2006). LakeSuperiorStreams. org: Making stormwater and stream data come alive for citizens, students, teachers, contractors, resource agencies, decision-makers and scientists. Presented at the 2006 National Water Quality Monitoring Conference, San Jose, CA May 7–11, 2006. Retrieved from www.lakesuperiorstreams.org/general/articles/AxlerDuluthStreamsNWQMC5-11-06.pdf

Barnhill, K. & Smardon, R. (2012). Gaining ground: Green infrastructure attitudes and perceptions from stakeholders in Syracuse, New York. *Environmental Practice*, 14, 6–16. doi: 10.10170S1466046611000470.

Bos, D.G. & Brown, H.L. (2015). Overcoming barriers to community participation in a catchment-scale experiment: Building trust and changing behavior. *Freshwater Science*, 34(3), 1169–1175. doi: 10.1086/682421

Center for Watershed Protection. (2004). *Urban Stream Repair Practices: Version 1.0. Urban Subwatershed Restoration Manual Series: Manual 4*. Ellicott City, MD: Center for Watershed Protection. Retrieved from https://owl.cwp.org/mdocs-posts/urban-subwatershed-restoration-manual-series-manual-4/

Cutts, B.B. (2013). Evaluating collective effects: A participatory approach to mapping public information about water issues in an uncertain and politicized context. In P. Lawrence (Ed.), *Geospatial Tools for Urban Water Resources* (pp. 37–60). New York, NY: Springer Press, NY.

Dernoga, M.A., Wilson, S., Jaing, C., & Tutman, F. (2015). Environmental justice disparities in Maryland's watershed restoration programs. *Environmental Science & Policy 45*, 67–78. doi: 10.1016/j.envsci.2014.08.007

Federal Interagency Stream Restoration Working Group (FISRWG). (1998). *Stream Corridor Restoration: Principles, Process, and Practices*. Washington, DC: Federal Interagency Stream Restoration Working Group, National Technical Information Service. Retrieved from www.nrcs.usda.gov/wps/portal/nrcs/detailfull/national/water/manage/restoration/?cid=stelprdb1043244

Findlay, S.J. & Taylor, M.P. (2006). Why rehabilitate urban river systems? *Area* 38(3), 312–325. doi: 10.1111/j.1475-4762.2006.00696.x

Ghaemi, P., Swift, J., Sister, C., Wilson, J.P. & Wolch, J. (2009). Design and implementation of a web-based platform to support interactive environmental planning. *Computers, Environment and Urban Systems*, 33(6), 482–491. doi: 10.1016/j.compenvurbsys.2009.05.002.

Guzy, M.R., Smith, C.L., Bolte, J.P., Hulse, D.W. & Gregory, S.V. (2008). Policy research using agent based modeling to assess future impacts of urban expansion into farmlands and forests. *Ecology and Society*, *13*(1). Retrieved from www.ecologyandsociety.org/vol13/iss1/art37/

Harrison, J., Ebert, D., Wade, T. & Yankee, D. (n.d.). Using ATtILA (Analytical Tools Interface for Landscape Assessments) to estimate landscape indicators and target restoration needs. Retrieved from www.academia.edu/23635990/Integrating_a_Landscape_Hydrologic_Analysis_for_Watershed_Assessment

Herringshaw, C.J., Thompson, J.R. & Stewart, T.W. (2010). Learning about restoration of urban ecosystems: A case study integrating public participation, stormwater management, and ecological research. *Urban Ecosystems*, *13*, 535–562. doi: 10.1007/s11252-010-0134-

Hulse, D.W., Branscomb, A. & Payne, S.G. (2004). Envisioning alternatives: Using citizen guidance to map future land and water use. *Ecological Applications*, *14*(2), 325–341. doi: 10.1890/02-5260

Im, W. (2011). Volunteer monitoring using community participatory mapping. Abstract in 17th *Annual WWMC Conference*, North Linthicum, MD, 37.

Kloss, C., Cararusse, C. & Stoner, N. (2006). *Rooftops to Rivers: Green Strategies for Controlling Stormwater and Combined Sewer Overflows*. New York, NY: Natural Resources Defense Council. Retrieved from www.chs.ubc.ca/archives/files/rooftops-to-rivers.pdf

Kondolf, G.M., Boulton, A.J., O'Daniel, S., Poole, G.C., Rahel, F.J., Stanley, E. H., Wohl, E., Bang, A., Carlstrom, J., Cristoni, C., Huber, H., Koljonen, S., Louhi, P. & Nakamura, K. (2006). Process-based ecological river restoration: Visualizing three-dimensional connectivity and dynamic vectors to recover lost linkages. *Ecology and Society*, *11*(2). Retrieved from www.ecologyandsociety.org/vol11/iss2/art5/

Leach, W.D., Pelkey, N.W. & Sabatier, P.A. (2002). Stakeholder partnerships as collaborative policymaking: Evaluation criteria applied to watershed management in California and Washington. *Journal of Policy Analysis and Management*, *21*(4), 645–670. doi: 10.1002/pam.10079

Montalto, F., Bartrand, T.A., Waldman, A.M., Travaline, K.A., Loomis, C.H., …Boles, M. (2013). Decentralized green infrastructure: The importance of stakeholder behavior in determining spatial and temporal outcomes. *Structure Infrastructure Engineering: Maintenance, Management, Life-cycle Design and Performance*, *9*(12), 1187–1205. doi: 10.1080/15732479.2012.671834National

Moran, S., Perreault, M. & Smardon, R. (2013). Finding our way: Urban waterway restoration and participation process. In J.G. Fabos, M. Lindhult, R.L. Ryan, & M. Jackson (Eds.), *Proceedings of Fabos Conference on Landscape and Greenway Planning: Pathways to Sustainability*. Amherst, MA: University of Massachusetts.

Nassauer, J.I., Kosek, S.E. & Corry, R.C. (2001). Meeting public expectations with ecological innovation in riparian landscapes. *Journal of the American Water Resources Association,* *37*(6), 1439–1443. doi: 10.1111/j.1752-1688.2001.tb03650.x

Onondaga Environmental Institute (OEI). (2009). *Onondaga Creek Conceptual Revitalization Plan*. Syracuse, NY: Onondaga Environmental Institute. Retrieved from www.oei2.org/OEIResources_OCRPDRAFT.html

Özgüner H., Eraslan, Ş. & Yilmaz, S. (2012). Public perception of landscape restoration along a degraded urban streamside. *Land Degradation and Development*, *23*(1), 24–33. doi: 10.1002/ldr.1043

Perreault, T., Wraight, S. & Perreault, M. (2012). Environmental injustice in the Onondaga Lake waterscape, New York State, USA. *Water Alternatives*, 5(4), 485–506.

Petts, J. (2006). Managing public engagement to optimize learning: Reflections from urban river restoration. *Human Ecology Review*, 13(2), 172–181.

Petts, J. (2007). Learning about learning: Lessons from public engagement and deliberation in urban river restoration. *The Geographical Journal*, 173(4), 300–311. doi: 10.1111/j.1475-4959.2007.00254.x

Pinkham, R. (2000). *Daylighting: New Life for Buried Streams*. Old Snowmass, CO: Rocky Mountain Institute.

Rhoads, B.L., Urban, M. & Herricks, E.E. (1999). Interaction between scientists and nonscientists in community-based watershed management: Emergence of the concept of stream naturalization. *Environmental Management 24* (3): 297–308.

Riley, A. (2016). *Restoring Neighborhood Streams*. Washington, D.C.: Island Press.

Rivera, S., Band, L.E., Lee, J.S., McHenry, K., Schmidt, A.R., Snoeyink, J., Sullivan, W.C., Whitton, M.C. & Minsker, B. (2014). Proposing a framework for crowd-sourced green infrastructure design. In D.P. Ames, N.W.T. Quinn & A.E. Rizzoli (Eds.), *International Environmental Modelling and Software Society (iEMSs): 7th Intl. Congress on Env. Modelling and Software*. San Diego, CA.

Riverson J., Zhen, J., Shoemaker, L. & Fu-hsiung Lai. (2004) Design of a decision support system for selection and placement of BMPs in urban watersheds. *Critical Transitions in Water and Environmental Resources Management*, 1–10 doi: 10.1061/40737(2004): 40.

Sabatier, P.A., Foght, W., Lubell, M., Trachtenberg, Z., Vedlitz, A. & Matlock, M. (2005). *Swimming Upstream: Collaborative Approaches to Watershed Management*. The MIT Press, Cambridge MA.

Smardon, R., Felleman, J. & Senecah, S. (1996). *Protecting Floodplain Resources: A Guidebook for Communities*. Federal Emergency Management Agency publication 268. Federal Interagency Floodplain Management Task Force, Washington, D.C.

Souder, J. (2013). The human dimensions of stream restoration: Working with diverse partners to develop and implement restoration. In P. Roni and T. Beechi (eds.) *Stream and Watershed Restoration; A Guide to Restoring Riverine Processes and Habitats* (pp. 114–143). Oxford, U.K.: Wiley-Blackwell

Tzoulas, K., Korpela, K., Venn, S., Yli-Pelkonen, V., Kazmierczak, A., ... James, P. (2007). Promoting ecosystem and human health in urban areas using green infrastructure: A literature review. *Landscape and Urban Planning*, 81, 167–178. doi: 10.1016/j.landurbplan.2007.02.001.

USEPA. (2002). *Community Culture and the Environment: A Guide to Understanding Sense of Place*. EPA 842-B-01-003. U.S. Environmental Protection Agency, Office of Water, Washington, D.C.

Woolsey S., Capelli, F., Gosner, T., Hoehn, E., Hostmann, M., ... Peter, A. (2007). A strategy to assess river restoration success. *Freshwater Biology (52):* 752–769 doi: 10.1111/j.1365-2427.2007.01740.x

Zidar, K., Bartrand, T.A., Loomis, C.H., McAfee, C.A., Geldi, J.M., ... Montalto, F. (2017). Maximizing green infrastructure in a Philadelphia neighborhood. *Urban Planning* 2(4): 115–132 doi 10.176545/up.v.v2i4.1039.

8 Urban waterways as green infrastructure

Richard Smardon and April Karen Baptiste

Introduction

In this chapter we will cover how different components of green infrastructure (GI) project management relate to environmental justice (EJ) issues. Some of the key questions that we raise are: In what ways does assessing benefits of restoration using standard economic measures fall short of EJ goals? In what ways does GI implementation reinforce existing dynamics and values versus transforming them?

The physical impacts of urbanization on stream ecosystems are covered in Chapters 1 and 2, but in summary here they include:

- Increase in impervious surface cover, which causes alteration of stream hydrology and geomorphology and changes in stream habitat (Gergel et al., 2002).
- Runoff from urbanized surfaces plus municipal and industrial discharges, causing increased loading of nutrients, metals, pesticides and other contaminants (Gergel et al., 2002).
- The first two points lead to decreased richness of algae, invertebrates and fish communities (Brown et al., 2005; Paul & Meyer, 2001; Violin et al., 2011).

From Chapters 3 and 4 we draw from the principles of distributive and procedural environmental justice issues. Are some communities benefiting from GI implementation and other communities not (distributive)? Are some voices heard as part of the process and other communities' voices not heard (procedural)?

There are physical health, social, psychological and equity issues for communities located along degraded urban waterways (Moran, 2010; Tzoulas et al., 2007). In this chapter we are interested in the use of revitalized urban creeks and streams as a way of providing urban open space, improved public health and remedying past environmental justice issues (Moran, 2010; Perreault et al., 2012; Pincetl & Gearn, 2005; Tzoulas et al., 2007; Wolch et al., 2014; Yli-Pelkonen et al., 2006).

There is also international interest in utilizing urban waterway revitalization as a means of creating more sustainable urban communities; some examples from the literature are located in Australia (Findlay & Taylor, 2006), Finland (Yli-Pelkonen et al., 2006), France (Dufour & Piegay, 2009), Korea (Nam-choon, 2005), Sweden (Andersson et al., 2015) and the U.K. (Everard & Moggridge, 2012), as well as the U.S. (Cairns & Palmer, 1995; Platt, 2006). We can also develop accounting systems for waterways serving as green infrastructure to deliver benefits to creekside communities in the form of ecosystem services (Everard & Moggridge, 2012).

The following sections of this chapter will cover a review of ecosystem services that urban waterways can provide with an emphasis on cultural services, ways of assessing ecosystems service and some case studies of applied ecosystem service assessment of benefits as well as environmental justice impacts to urban waterway communities. In addition, we will also explore green infrastructure using the lens of environmental justice, identifying several challenges and limitations that arise in connection with the context of restoration projects.

Review of urban waterway ecosystem services

The Millennium Ecosystem Assessment project (2005b) defines ecosystem services as the "benefits people obtain from ecosystems" (p. 3). These services include provisioning services, regulating, supporting and cultural services. For urban waterways, streams and creeks, *provisioning* services could include water supply and fish as food. *Regulating* services could include streamside wetlands reducing water pollution. *Support* services could include nutrient cycling. *Cultural* services could include recreational, spiritual and other nonmaterial benefits. For urban waterways we are particularly interested in those ecosystem services that provide well-being for urban residents, especially those they provide for health, good social relations and security.

Green infrastructure can be considered as all-natural, semi-natural and artificial networks of multifunctional ecological systems within, around and between urban areas, at all spatial scales (Tzoulas et al., 2007, p. 169). In this case we are interested in the ability of urban waterways to provide ecosystem services as green infrastructure. The GI examples shown in Figures 8.1, 8.2 and 8.3 are taken from the Bronx River, illustrating the use of riparian vegetation, shoreline anchoring and green roofs to reduce both storm runoff and erosion along the Bronx River.

One ambitious form of GI is reintroduction of stream-edge wetlands along urban streams to reduce sediment and nutrient loads as well as provide habitat and urban amenities. One of the best examples is the Des Plaines River wetland demonstration project (www.wetlandsresearch.org/projects) north of Chicago, which reduced much of the pollutant load of the river by running through six artificially re-created wetlands (Sather, 1992). In addition, specific wetlands have been designed to reduce the water flow

Figure 8.1 Example of Bronx River GI with riparian forest
Source: J. Weis

Figure 8.2 Example of reinforced shoreline GI
Source: J. Weis

Figure 8.3 Green roof adjacent to Bronx River
Source: J. Weiss

amount and pollution from combined sewer overflows (Levy et al., 2014; Tao et al., 2014).

An example of an early green infrastructure system along a creek would be the Back Bay fens in Boston, which are linked together in a greenway that provides both regulatory (clean air and carbon sequestration) and cultural (recreation) ecosystem services (Figure 8.4). The green infrastructure of the Boston fens includes the creekside wetlands that absorb some of the water contaminants and the parks along the fens where trees and other vegetation absorb some of the air pollutants. These parks also provide amenity and recreational benefits for Boston residents and visitors.

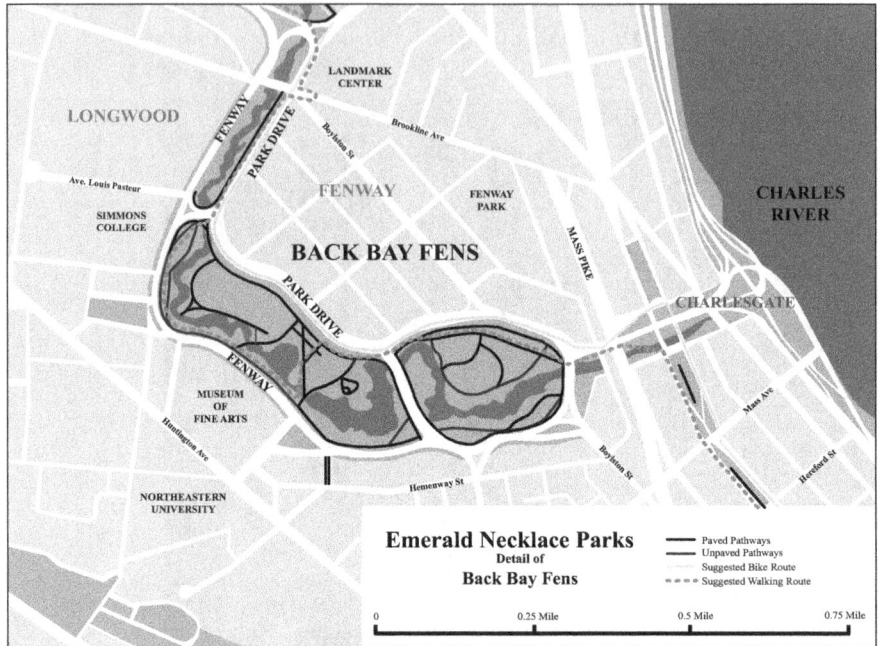

Figure 8.4 Map of the Back Bay fens
Source: Drawn by Ryan Mackerer

If one looks at the larger scale drainage basins one could use a range of GI measures to reduce urban runoff, including rainwater harvesting, rain gardens, bioswales, permeable pavement, green streets and parking areas, green roofs and urban tree canopy development (see USEPA, n.d. for much more GI information). In such large-scale drainage basins, GI has been implemented along urban rivers in Chicago, Cleveland, New York City, Philadelphia and Syracuse, NY.

Ways of assessing ecosystem services

Green infrastructure can be assessed in terms of supplying benefits to urban residents. These benefits can include providing both new open space opportunities and partial remediation of previous environmental (in) justice impacts. Assessing ecosystem services, especially cultural ecosystem services, can be a way of identifying and accounting for these benefits. There are multiple ways to assess ecosystem services. A few of these will be highlighted in this section. The United Kingdom approach (Defra, 2007; Everard & Moggridge, 2012; Everard, 2012) includes an impact pathway related to any policy change. Stepwise the pathway includes: Policy change > Impacts

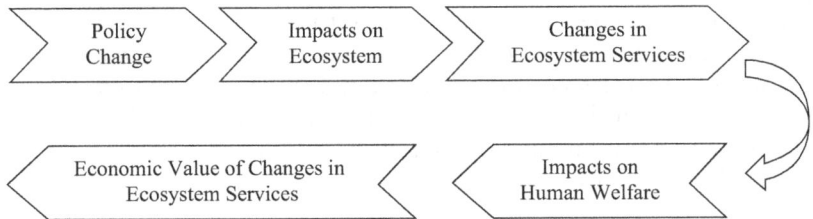

Figure 8.5 U.K. Pathway for policy changes

on ecosystem > Changes in ecosystem services > Impacts on human welfare > Economic value of changes in ecosystem services (Figure 8.5).

So, for example, if there is an urban waterway revitalization program that is implemented as a specific creek project > that causes a change in creek hydrology and ecology > that causes a change in local creek residents' health (pollution reduction), and so the linkages are documented in some way. The Defra (2007) guide provides a practical introduction to the key steps to be undertaken in valuing ecosystem services in a policy appraisal context. The guide takes an impact pathway approach to valuing ecosystem services. In summary, those key steps are:

1. Establish the environmental baseline.
2. Identify and provide qualitative assessment of the potential impacts of policy options on ecosystem services.
3. Quantify the impacts of policy options on specific ecosystem services;
4. Assess the effects on human welfare; and
5. Value the changes in ecosystem services.

Everard is one of the key architects of this system, which has been applied to the Mayes Brook in East London, and which will be reviewed as one of the case studies of GI implementation (Everard, 2012) as well as a number of other case studies (Everard, 2009, 2010).

The U.S. Environmental Protection Agency/National Research Council approach to ecosystem services (Landers & Nahlik, 2013; Ringold et al., 2013; USEPA, 2009) has been developed for streams and rivers as well as wetlands and estuaries. This system differentiates between intermediate and final ecosystem services—in other words, the ability of a streamside wetland to improve water quality, which then results in a final service of improving human and ecosystem health. This system has resulted in a classification system (Landers & Nahlik, 2013) for major types of environments delivering ecosystem services that include rivers and streams, wetlands, lakes and ponds, estuaries, open oceans and seas.

An example of a cultural service within the rivers and streams matrix includes the recreational experiences of viewers. The general beneficiary description includes "beneficiary views and experiences the environment via an activity, such as scenery gazing, hiking, bird watching, botanizing, ice skating, rock climbing, flying kites, etc. This beneficiary has potential contact with water" (Landers & Nahlik, 2013, p. 48). Final Ecosystem Goods and Services (FEGS) in this case could include the presence of the environment, viewscapes, flora, fauna, sounds and scents. Importance of the FEGS to the beneficiary includes:

- Opportunity to view the environment and the organisms within it.
- Landscape that provides a sensory experience.
- Organisms (e.g., flowers and plants) that can be viewed.
- Organisms (e.g., birds, mammals and reptiles) that can be viewed; and
- Sounds that provide a sensory experience (Landers & Nahlik, 2013, p. 48).

The development of potential metric(s) and indicator(s) is in progress by the USEPA Ecosystems Laboratory in Corvallis, Oregon.

There also has been development of ecosystem services for wetlands (Millennium Ecosystem Assessment, 2005a; Maltby & Acreman, 2011) along waterways. Smardon (1983) developed one of the earliest cultural resource assessment systems for freshwater wetlands in the northeastern U.S. that focused on recreation, aesthetic and educational values of wetlands. The Millennium wetland ecosystem service assessment system (MEA, 2005a) focuses on the well-being and livelihood benefits that can be gained from wetlands worldwide.

Scale- and context-dependent issues for multifunctional landscapes are addressed by Andersson et al. (2015) to meet the needs of expanding urban populations. Their work provides the concept of service providing units (SPUs) as a way to plan and manage structures and preconditions that are needed for or influence providing ecosystems services. The SPU approach has two parts; the first addresses the internal dimensions of the SPUs, and the second looks at how the context and presence of external structures such as built infrastructure and larger ecosystem affect the performance of SPUs in delivering ecosystem services.

One of the major issues for urban water investment and green infrastructure development is how ecosystem services are used within the overall cost benefit analysis. Some argue that traditional cost benefit analysis alone justifies development of green infrastructure for water quality treatment and urban runoff reduction (Jaffe, 2011). Others argue that use of an integrated water resource management with ecosystem services can effectively facilitate systematic consideration and quantitative assessment of broad impacts of water infrastructure development (Kondulu et al., 2014). In Adelaide, Australia, five ecosystem services were quantified, including urban recreational amenities, regulation of coastal water quality, salinity, greenhouse gas emissions and support of estuarine habitat.

The other issue often affecting smaller urban waterways is the willingness of local residents to accept implementation of green infrastructure as well as to pay for it to gain ecosystem services such as urban runoff reduction (Bae, 2011; Baptiste et al., 2015). Bae (2011) has studied willingness to pay for restoration process of urban waterways in Korea. He has found that transforming existing concrete-encased streams into natural state streams increases the value of an urban stream by about US$50 per household, and improving recreational attributes of an urban stream (i.e., walkway installation) increases the value of an urban stream by about US$25 per household. In Syracuse, New York, USA we looked at residents' willingness to implement green infrastructure to reduce urban storm runoff and found that key factors were efficacy, aesthetics and cost (Baptiste et al., 2015). We also found that urban residents often do not understand what ecosystem services are (Barnhill & Smardon, 2012).

In summary, the limitations of a purely economic assessment of ecosystem services are:

- Unquantifiable values;
- Double counting and overlap of recreational, aesthetic and educational experience while boating down an urban creek;
- Tradeoffs between benefits in restoration decision-making, e.g., ecosystem improvement, versus urban resident safety or amenity values, which is an EJ substantive issue;
- Lack of engagement of urban residents, which is an EJ procedural issue;
- Lack of consideration of ethical issues in general, which also aligns with EJ principles along the lines of the right to good quality environment; and
- Spatial scale relationship to beneficiaries, e.g., one creek stretch benefits while another downstream does not or sees fewer benefits, which is an EJ distributional issue.

On the other hand, assessing cultural ecosystem services has its own challenges, which include:

- Measuring or assessing intangible values such as the inspirational value of a historical event on a waterway;
- Complexity in assessing overlapping cultural ecosystem services such as amenity values, e.g., aesthetic, educational and inspirational values of a historic waterway such as the Boston Fens;
- Fluctuating value attribution or perception, e.g., what was once perceived as scenic waterway may change over time;
- Difficulty in geographic attribution, such as a historic event along a waterway that is difficult to tie down to a specific creek stretch; and
- Inconsistent valuation where willingness to pay for a benefit may change over time.

So the issue is how can we use ecosystem service assessment and other valuation systems to address urban waterway benefits regarding human health and access to open space, as well as address equity EJ issues? The following case studies may help.

Ecosystem service case studies

Mayes Brook, East London, case study (Everard, 2011)

The Mayes Brook runs through the London Borough of Barking and Dagenham (LBBD), which comprises 17 wards, of which 5 are ranked within the 10 percent most deprived wards in England and a total of 14 wards are ranked within the 20 percent most deprived. Socio-economic statistics indicate that the borough has a relatively high unemployment rate (5.4 percent) and low household incomes compared with London-wide and national levels. Health deprivation in the borough is also linked to high rates of teenage pregnancy, cancer and heart disease, and below-national and London average life expectancy (Everard, 2012).

Water quality is affected by several surface drains that discharge into the Mayes Brook and by large quantities of litter, which frequently accumulate on the screening, grills covering the larger inlets. The brook also receives effluent from many misconnections. Reported pollution incidents (mostly minor incidents) mainly occur in the northern lake, probably associated with storm drainage during high flow events. These incidents represent a significant pollutant load in addition to more general diffuse urban pollution.

Mayesbrook Park lies toward the middle section of the Mayes Brook catchment (see Figure 8.6). It covers an area of around 45 hectares. To the southern (downstream) end of the park lie two linked lakes, created as a result of sand and gravel extraction between 1919 and 1938 as London expanded. A decision was taken in the 1930s not to build on this area but to retain it as an urban park among the sprawl of development. The development of the park was interrupted in 1939 by the start of the Second World War, and the Italianate gardens and other features were never completed. The park is now surrounded by dense urban development, including many housing estates and associated infrastructure. The Upney Underground Station, roughly a kilometer from the southwest boundary of the park, provides a rail link to central London and eastward to Upminster. Around 1.6 km of the Mayes Brook runs through Mayesbrook Park, defining its border to the north and west. The brook is currently completely disconnected from the park. It is fenced off on the park side, and also largely invisible from the park as the channel is deeply sectioned. An embankment on the park side of the channel, resulting from an accumulation of spoils dredged from the brook and piled on the bank, further blocks the view and water flows between river and park. This potentially poses a flood risk to adjacent properties on the right bank, which lies at a lower level than the park side

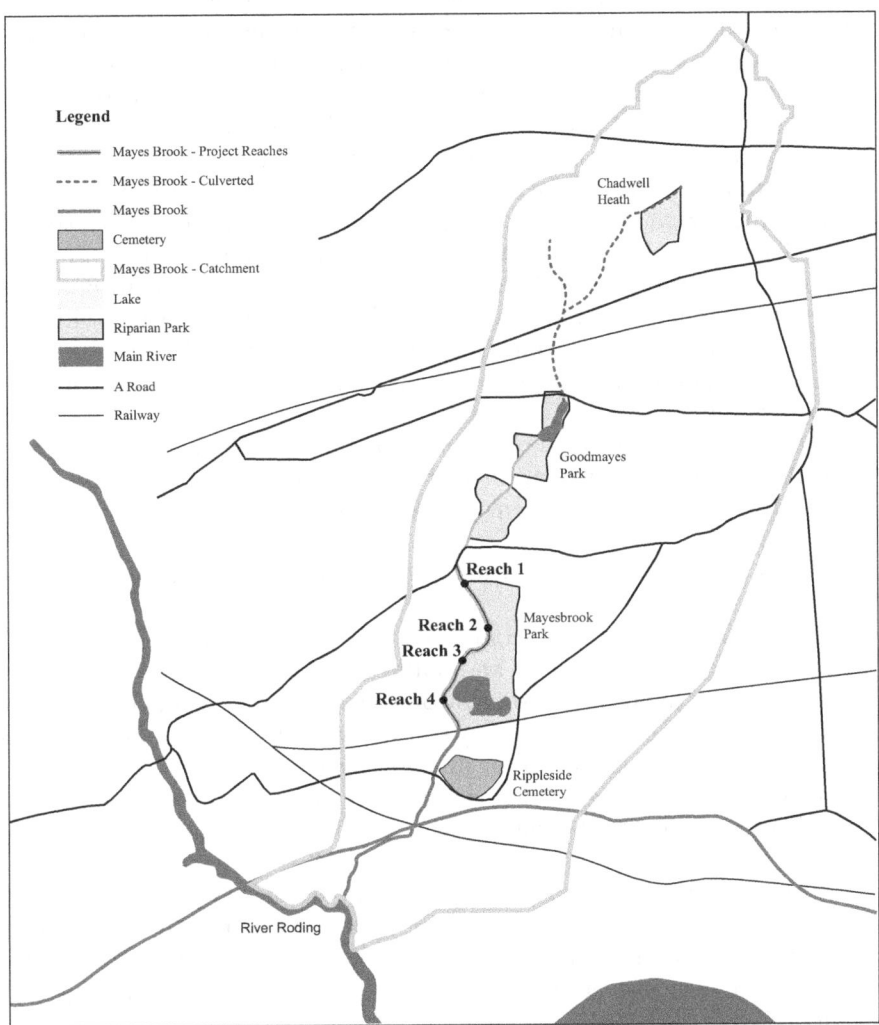

Figure 8.6 Map of Mayes Brook with proposed work reaches
Source: Drawn by Ryan Mackerer

embankment on the left bank. Furthermore, along the right bank top and face, several of the adjoining residential gardens have encroached onto the two-to-three-meter right bank, which now includes areas of planting as well as dumping sites for garden waste.

Today, much of the park area is covered by short mown grass, which provides poor habitat for wildlife and is not used intensively by the neighboring community. Around the two lakes (Figure 8.6), large quantities of feces produced by native and non-native birds (predominantly Canada geese) present a health risk. The lakes have become heavily polluted as they have served as a sink for substances from the Mayes Brook stormwater overflow entering via the connecting high flow inlet channel. Three meters of polluted sediment have now accumulated in the top lake. The top lake has an overflow to the bottom lake, and there is a further outflow from the bottom lake into the Mayes Brook immediately before the brook flows through the screened entrance to the culvert downstream of Mayesbrook Park (Figure 8.7a and b). Fishing and boating were historically popular activities on the lakes. Both have now been stopped due to pollution concerns, although kayaking continues when conditions permit, with regular sessions run by the Barking and Dagenham Canoe Club on the upper lake, despite pollution and interruptions caused by unsafe levels of blue-green algal blooms.

The case sets out a range of options for further enhancing public value from the restoration scheme, through new or redesigned initiatives or in green infrastructure management practices. These include:

- Enhancing the hydrological function of the whole park landscape and infrastructure;
- Using reed bed filtration to improve water quality in a bypassed reach of river and at lake inflows and outflows;
- Improving climate regulation through energy-efficient building design, installation of renewable energy sources and reusing tree and other park trimmings as biomass fuel (or mulch) on site; and
- Optimizing park restoration design to provide health and educational resources to the local community.

Assessing the ecosystem service implications for these options and others that may be identified in later phases of planning and research would help support the economic case for their implementation. The result of the ecosystem assessment of the green infrastructure benefits is displayed in Figure 8.8.

The most notable benefits of the Mayesbrook Park restoration come from cultural services to creekside communities, which account for around 93 percent of total benefits (40-year net present value [NPV] of annual benefits + 100-year property uplift). Overall benefits are substantial relative to investment, representing a lifetime benefit-to-cost ratio of around 7:1. The urban setting and impoverished biodiversity of Mayesbrook Park and the stretch of the Mayes Brook that runs through it mean that restoration will bring no further benefits from provisioning services. This is a different outcome from other assessments in this series of Environment Agency case studies of interventions in the aquatic environment, all of which have addressed rural areas in the U.K.

Figure 8.7a Mayesbrook Park activity
Source: Photo by John Sutton at www.geogrph.org.uk

Figure 8.7b Mayesbrook Park pond
Source: Dudley Miles (own work) [CC BY-SA 3.0 (https://creativecommons.org/licenses/by-sa/3.0)], via Wikimedia Commons

Ecosystem Service	Annual Benefit Assessed Research Gap/Note
Gross annual provisioning service benefits	**There is no increase to provisioning services.** This contrasts markedly with related rural case studies (Everard, 2009a and 2010; Everard and Kataria, 2010), where impacts on farm profits significantly affect this service category. Some development options (reuse of trimmings for 'fiber and fuel') may potentially produce provisioning service benefits.
Gross annual regulatory service benefits	**Gross annual regulatory service benefits are approximately £ 28,000** (calculated total = £28,087) comprising climate regulation at £13,000 + flood risk at £10,000 + erosion at £5,000. However, there will also be **'likely significant positive benefits'** for the regulation of air quality and microclimate. All of these benefits relate almost entirely to public health and risk management, showing the potential role of Mayesbrook Park to enhance the well being of the neighborhood.
Gross annual cultural service benefits	**Gross annual cultural service benefits are approximately £820,000** (calculated total = £820,169) comprising recreation and tourism at £815,000 + educational value at £5,000. However, the net uplift (via 'social relations') to **regional regeneration is assessed with a lifetime (100 year) benefit of at £7,800,000** which will be factored into the final NPV calculation.
Gross annual supporting service benefits	**Gross annual supporting service benefits are approximately £31,000** (calculated total = £30,573) comprising nutrient cycling at £21,000 + habitat for wildlife at £10,000.
Total annual ecosystem services uplift across the four categories	**Gross annual ecosystem service benefits are approximately £880,000** (total = £878,829 based on summing calculated values to avoid rounding errors) but there are also **'likely significant positive benefits'** for the regulation of air quality and microclimate as well as a (100-year) contribution to **regional regeneration of £7,800,000.**

Figure 8.8 Ecosystem calculations

Source: Redrawn by Ryan Mackerer and adapted from Everard (2012)

In contrast, the urban setting means that ecosystem enhancements can make major contributions to regulatory services (regulation of air and water quality, microclimate and flood risk) affecting the local community. The same is true for cultural services (recreation and tourism, social cohesion and educational opportunities), particularly since many people in the borough lack gardens or ready access to other green spaces.

Supporting services, which are hard to quantify but essential for maintaining ecosystem functions underpinning more directly used services, are significant in terms of nutrient cycling and provision of habitat for wildlife. This habitat improvement helps ensure there are animals and plants capable of colonizing the wider landscape as the habitat improves, also serving as a "stepping stone" for wildlife to move across and between limited and fragmented suitable habitat in the urban landscape.

This analysis suggests that much public value will accrue from urban river and parkland restoration, fully justifying the planned investment and providing firm evidence that investment in urban green infrastructure is highly favorable to the health and well-being of local people and economic uplift of deprived wards. This scheme is a cost-effective means of improving well-being and quality of life in urban communities. There also would be potential for applying the principles of environmental justice to this case.

Rekolanoja, Finland, case study (Yli-Pelkonen et al., 2006)

> The stream Rekolanoja is located mainly in the City of Vantaa ... although the catchment area of the stream (37.75 km²) is partly located in the City of Kerava and the Tuusula municipality, where the stream originates. In this study, Rekolanoja is examined principally in the City of Vantaa, but land use of the entire catchment area affects the studied part of the stream. Rekolanoja flows southward into the Keravanjoki River, which in turn flows into the Vantaa River, which eventually flows into the Gulf of Finland, Baltic Sea [see Figure 8.9]. As is typical for urban streams, Rekolanoja experiences large fluctuation extremes because runoff waters flow rapidly into the stream. During high-water levels, the stream can attain widths of several meters.
>
> (Yli-Pelkonen et al., 2006, p. 675)

Today, the catchment area is intensively built up (48 percent built-up areas), mostly near the railway and the stream. Another 37 percent of the catchment

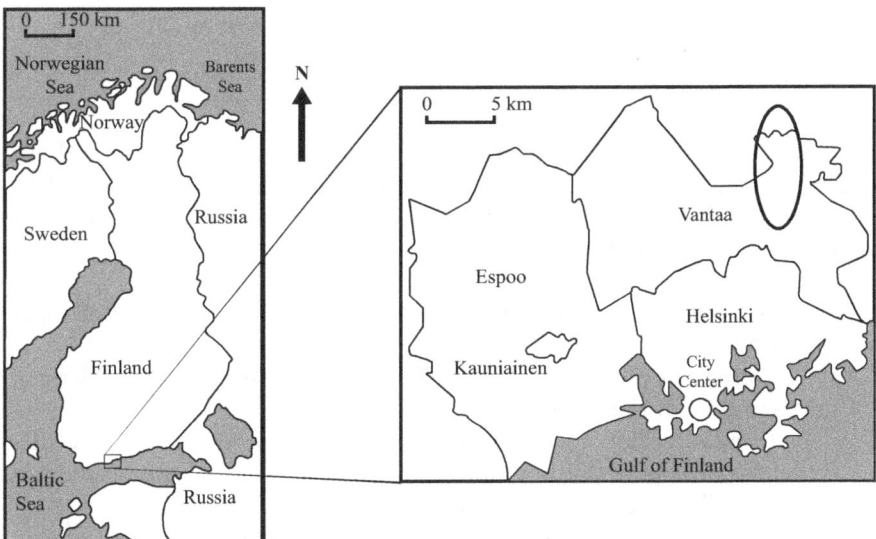

Figure 8.9 Rekolanoja catchment location
Source: Redrawn by Ryan Mackerer and adapted from Yi-Polkonen et al. (2006)

area constitutes forest, 9 percent fields and 6 percent bogs. Planned new residential districts and a commercial center will considerably increase the future proportion of built-up areas, especially near the railway and the stream. The importance of Rekolanoja and its streamside surroundings has been noted in the Green Area Program of the City of Vantaa. In the upstream City of Kerava, current planning policy states that in new residential areas the stream should also be treated as a natural landscape element instead of an element to be moved or piped. The planned new residential and commercial districts will put extensive pressure on the ecological resilience of the Rekolanoja stream ecosystem.

Study methods

> For purposes of the study [see Figure 8.10], the following methods were used: review of ecological studies and inventories, semi-structured interviews, a resident inquiry and a writing contest. Once the importance of the biodiversity of the stream had been studied based on the existing ecological data, the social importance of the stream and the linkages between human health, social well-being and the biodiversity of the stream ecosystem were addressed. Relevant planning and environmental officials, representatives of local resident and environmental associations and local residents and entrepreneurs related to the stream were interviewed. A resident inquiry and a writing contest were targeted to give more information on the ecological and social importance of Rekolanoja.
>
> (Yli-Pelkonen et al., 2006, p. 677)

Results

All respondents to the resident inquiry and several interviewed local residents noted that the stream environment is appealing and affects their quality of life in a positive way. The quotes below depict some of the ways local residents perceive the stream:

> The stream makes the area feel less urban. I like the seasonal changes of the stream.
>
> The stream is pleasant, enlivens the landscape and is a water element in the middle of the urban landscape.
>
> Nature, water flowing freely, birds singing. In the springtime I wonder if nightingales arrive, although the stream and its sides have been excavated heavily – luckily have arrived so far.
>
> The streamside bushes have been heavily reduced, but the stream should be preserved as naturally as possible. Well-being increases from nature and natural things.
>
> (Yli-Pelkonen et al., 2006, p. 680)

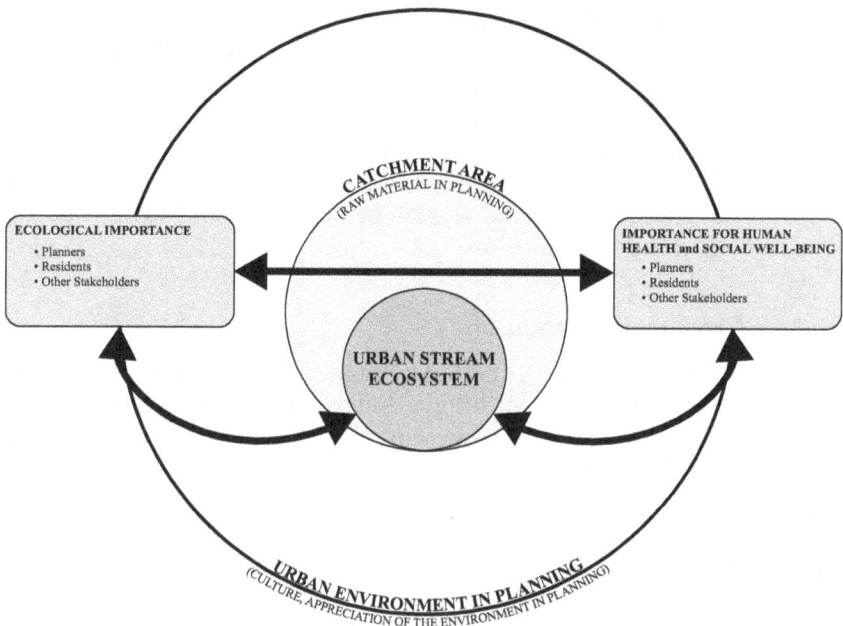

Figure 8.10 Conceptual framework
Source: Redrawn by Ryan Mackerer and adapted from Yi-Polkonen et al. (2006)

Comments related to the stream indicate that it has helped shape the local identity in the Rekolanoja area. The interviews and the writing contest indicated that the stream used to be a recreational oasis for local dwellers in the 1930s to 1950s, providing opportunities for fishing, crayfishing, swimming and children's play. The memories from the 1960s, by contrast, were associated with waste in the stream and damage to grove meadows.

The recreational value of the stream corridor is important for several reasons. The water element as such, including the sounds of the stream and the singing of birds, is appealing and calming. The stream was historically regarded as a desirable environment offering silence, ease and relaxation away from urban noises and stress. For some, unhurried walking and tranquility on the streamside serve as recreation, and the paths on the streamside offer opportunities for dog walking or jogging and observing the local surroundings at the same time. For many residents and recreational users the stream and the streamside green space provide aesthetic values, and for some its mere existence is important. However, several interviewees raised the issue of the untidiness of the stream and stream banks.

The interviews and resident inquiries showed that the stream is also an important part of the environment for teachers and pupils in local schools.

The interviewees and respondents hoped that schools in the area would use the stream increasingly in environmental education. All the interviewed residents agreed that the ecological values of the stream should be preserved. However, local residents have conflicting desires. While some hoped for more unmanaged green areas by the streamside, others wanted more park-like green areas. In situations where there may be conflicts over how the resource should be used, principles of procedural justice must be considered, i.e., all the stakeholders involved should be consulted and their concerns must be valued in the final decision.

Onondaga Creek revitalization case

Syracuse, New York, is near the end of a massive cleanup of the Onondaga Lake watershed, including a half-billion-dollar effort to upgrade to tertiary treatment of the main sewage treatment plant emptying into Onondaga Lake, with another half-billion dollars to remediate mercury-contaminated sediment in the bottom of the same lake. As part of the water quality cleanup there is the issue of combined sewerage overflows (CSO) along Onondaga Creek that dump raw sewage plus street drainage any time there is a major rainstorm. This is typical of many northeastern and Great Lakes urbanized areas. EPA has called CSOs "remnants of the country's early infrastructure" (USEPA, 2011, p. 2). Combined sewer systems serve upward of 772 communities nationwide, which are home to more than 40 million people (USEPA, 2011).

The Syracuse area has had a history of "redlining" for the lower-income residents living along the creek on the City's Southside who are subjected to both the CSOs and major floods. As a result of this flooding the perceived solution along the creek's length within the city was channelization, which results in massive flow rates during major storm events. The creek within the city limits is fenced off for safety reasons related to storm events, thus limiting greenway access.

The upper middle portion of Onondaga Creek flows through the Onondaga Nation, and the whole watershed is part of their original territory and is sacred to them. Both the Southside and the Onondaga Nation's historical significance constitute environmental justice issues, which are described in detail by Perreault et al. (2012).

For the Native American community, the Onondagas, this was one of the very few planning exercises they fully participated in. The degree of inclusion of the Onondaga Nation is documented in a master's thesis by Barnhill (2009). In terms of process, we extensively utilized co-production with the working group and the project team throughout the planning process. This co-production is emphasized by other work in Europe (Petts, 2006, 2007) as well as in North America (Daniels & Walker, 1996; Lubell et al., 2002; McGinnis, 1999; Smolko et al., 2002; Walker Senecah, & Daniels, 2006). From an environmental justice perspective, such a process is critical in overcoming earlier real and perceived injury to the communities previously impacted (Gottlieb, 2009; Light & Higgs, 1996; Moran, 2010; Perreault et al., 2012; Riley, 1989).

Within the city of Syracuse many of the CSOs occurred in the Southside neighborhood, thus exposing those residents to water pollution during storm events. The perceived EJ issue was the siting of the Midland Avenue treatment facility in a Southside neighborhood. From an engineering point of view this made sense as an ideal collection point for CSO water treatment. From the neighborhood's point of view this was another industrial sewage treatment plant, further damaging the neighborhood's integrity. This case was forwarded to the U.S. Environmental Protection Agency as an environmental justice case. Later on green infrastructure was extensively implemented as the Save the Rain campaign within the city of Syracuse and surrounding areas to reduce the storm flow so that the combined sewer overflows (CSOs) were reduced, lowering the contaminates going into Onondaga Creek. So in the end the EJ issue impact on the Southside neighborhood was reduced by spreading green infrastructure to reduce storm flows, thus partially addressing a distributional EJ issue.

Summary

This chapter shows the potential of treating urban waterways as green infrastructure and providing ecosystem services for urban residents. Different methods of ecosystem service assessment are reviewed, giving a sense of how green infrastructure has been used to address some environmental injustices with respect to stream restoration. Finally, three case studies in the U.K., Finland and the U.S. are presented with specific reference to cultural ecosystem services as they affect human health and well-being.

References

Andersson, E., Tengo, M., McPhearson, T. & Kremer, P. (2015). Cultural ecosystem services as a gateway for improving urban sustainability. *Ecosystem Services, 12,* 165–168. doi: 10.101016/j.ecoser.2010.08.002

Bae, H. (2011). Urban stream consideration and residents' willingness to pay. *Urban Forestry and Urban Greening, 10,* 119–126. doi: 10.1016/j.ufug.2011.02.001

Baptiste, A.K., Foley, C., & Smardon, R.C. (2015). Understanding urban neighborhood differences in willingness to implement green infrastructure: A case study of Syracuse, NY. *Landscape and Urban Planning, 136,* 1–12. doi: 10.1016/j.landurbplan.2014.11.012

Barnhill, S. K. (2009). Negotiating sacred space: indigenous participation in local environmental governance. Unpublished Master's thesis. Syracuse, NY: SUNY College of Environmental Science and Forestry.

Barnhill, K. & Smardon, R. (2012). Gaining ground: Green infrastructure attitudes and perceptions from stakeholders in Syracuse, NY. *Environmental Practice, 14*(1), 6–16. doi: 10; 1017/s1466046611000470

Brown, L.R., Gray, R.H., Hughs, R.M. & Meader, M.R. (2005). Introduction to effects of urbanization on stream ecosystems. *American Fisheries Society Symposium, 47*, 1–8. Retrieved from http://citeseerx.ist.psu.edu/viewdoc/download? doi=10.1.1.482.7284&rep=rep1&type=pdf

Cairns, J. & Palmer, S.E. (1995). Restoration of urban waterways and vacant areas: The first steps toward sustainability. *Environmental Health Perspectives, 103*(5), 452–453. doi: 10.2307/3432582

Daniels, S.E. & Walker, G.B. (1996). Collaborative learning: Improving public deliberation in ecosystem-based management. *Environmental Impact Assessment Review, 18*(2), 167–174. doi: 10.1016/0195-9255(96)00003-0

Defra. (2007). *An Introductory Guide to Valuing Ecosystem Services*. London: Department for Environment, Food & Rural Affairs. Retrieved from http://ec.europa. eu/environment/nature/biodiversity/economics/pdf/valuing_ecosystems.pdf

Dufour, S. & Piegay, H. (2009). From the myth of a lost paradise to targeted river restoration: Forget natural reference and focus on human benefits. *River Research and Applications, 25*, 568–581. doi: 10.1002/rra.1239

Everard, M. (2012). What have rivers ever done for us? Ecosystem services and river systems. In B.J. Boon & P.J. Raven (Eds.). *River Conservation and Management* (pp. 313–321). London: Wiley-Blackwell.

Everard, M. (2011). *The Mayes Brook Restoration in Mayes Brook Park, East London: An Ecosystem Services Assessment*. Bristol, U.K.: Environment Agency. Retrieved from http://catalog.ipbes.net/system/assessment/193/references/files/568/original/ The_Mayes_Brook_restoration_in_Mayesbrook_Park_East_London_-_an_ ecosystem_service_assessment.pdf?1364316259

Everard, M. (2009). *Ecosystem Services Case Studies*. Bristol, U.K.: Environment Agency. Retrieved from http://catalog.ipbes.net/system/assessment/194/references/ files/569/original/Using_science_to_create_a_better_place_-_ecosystem_services_ case_studies_2009.pdf?1364317641

Everard, M. (2010). *Ecosystem Services Assessment of Sea Trout Restoration Work on the River Glaven, North Norfolk*. Bristol, U.K.: Environment Agency. Retrieved from www.gov.uk/government/uploads/system/uploads/attachment_ data/file/291657/scho0110brua-e-e.pdf

Everard, M. & Moggridge, H.L. (2012). Rediscovering the value of urban rivers. *Urban Ecosystems, 15*, 293–314. doi:10.107/s11252-011-0174-7

Findlay, S.J. & Taylor, M.P. (2006). Why rehabilitate urban river systems? *Area, 3*, 312–325. doi: 10.1111/j.1475-4762.2006.00696.x

Gergel, S.E., Turner, M.G., Miller, J.R., Mclack, J.M. & Stanley, E.H. (2002). Landscape indicators of human impacts to riverine systems. *Aquatic Sciences, 64*, 118–128. doi: 10.1007/s00027-002-8060-2

Gottlieb, R. (2009). Where we live, work, play and eat: Expanding the environmental justice agenda. *Environmental Justice, 2*(1), 7–8. doi: 10.1089/env.2009.0001

Jaffe, M. (2011). Reflections on green infrastructure economics. *Environmental Practice, 12*(4), 357–365. doi: 10.1017/S1466046610000475

Kondulu, J.M., Conner, J.D. & McDonald, P.H. (2014). Ecosystem services in urban water investment. *Journal of Environmental Management, 145*, 43–53. doi: 10.1016/j.jenvman.2014.05.024

Landers, D.H. & Nahlik, A.M. (2013). *Final Ecosystem Goods and Services Classification System (FEGS-CS)*. Washington, D.C.: USEPA Office of Research and Development. Retrieved from https://gispub4.epa.gov/FEGS/FEGS-CS%20 FINAL%20V.2.8a.pdf

Levy, Z.F., Smardon, R.C., Bays, J.S. & Meyer, D. (2014). A point source of a different color: Identifying a gap in United States regulatory policy for "green" CSO treatment using constructed wetlands. *Sustainability, 6,* 2392–2412. doi: 10.3390/su6052392

Light, A. & Higgs, E.S. (1996). The politics of restoration. *Environmental Ethics, 8,* 227–247. doi: 10.5840/enviroethics199618315

Lubell, M., Schneider, M., Scholz, J.T. & Metz, M. (2002). Watershed partnerships and the emergence of collective action situations. *American Journal of Political Science, 46*(1), 148–163. doi: 10.2307/3088419

Maltby, E. & Acreman, M.C. (2011). Ecosystem services of wetlands: Pathfinder for a new paradigm. *Hydrological Science Journal, 56*(8), 1341–1359. doi: 10.1080/02626667.2011.631014

McGinnis, M.V. (1999). Making the watershed connection. *Policy Studies Journal, 27*(3), 497–501. doi: 10.1111/j.1541-0072.1999.tb01982.x

Millennium Ecosystem Assessment. (2005a). *Ecosystems and Well Being: Wetlands and Water Synthesis.* Washington, D.C.: Water Resources Institute.

Millennium Ecosystem Assessment. (2005b). *Ecosystems and Human Well-Being: Our Human Plant: Summary for Decision Makers.* Washington, D.C.: Island Press.

Moran, S. (2010). Cities, creeks, and erasure: Stream restoration and environmental justice. *Environmental Justice, 3*(2), 61–69. doi: 10.1089/env.2009.0036

Nam-choon, K. (2005). Ecological restoration and revegetation works in Korea. *Landscape and Ecological Engineering, 1,* 77–83. doi: 10.1007/s1135-005-0011-3

Paul, M.J. & Meyer, J.L. (2001). Streams in the urban landscape. *Annual Review of Ecology and Systematics, 32,* 333–365. doi: 10.1146/annurev.ecolsys.32.081501.114040

Perreault, T., Wraight, S. & Perreault, M. (2012). Environmental injustice in the Onondaga Lake waterscape, New York State, USA. *Water Alternatives, 5*(4), 485–506.

Petts, J. (2006). Managing public engagement to optimize learning: Reflections from urban river restoration. *Human Ecology Review, 13*(2), 172–181.

Petts, J. (2007). Learning about learning: Lessons from public engagement and deliberation in urban river restoration. *The Geographical Journal, 173*(4), 300–311. doi: 10.1111/j.1475-4959.2007.00254.x

Pincetl, S. & Gearin, E. (2005). The reinvention of public open space. *Urban Geography, 26,* 365–381. doi: 10.2747/0272-3638.26.5.365

Platt, R.H. (2006). Urban watershed management: Sustainability one stream at a time. *Environment: Science and Policy for Sustainable Development, 48*(4), 26–42. doi: 10.3200/ENVT.48.4.26-42

Riley, A.L. (1989). Overcoming federal water policies: The Wildcat-San Pablo creeks case. *Environment: Science and Policy for Sustainable Development, 31*(10), 12–31. doi: 10.1080/00139157.1989.9928987

Ringold, P.L., Boyd, J., Landers, D. & Weber, M. (2013). What data should we collect? A framework for identifying indicators of ecosystem contribution to human well-being. *Frontiers in Ecology and the Environment, 11,* 98–105. doi: 10.1890/110156

Sather, H.J. (1992). *Intensive Studies of Wetlands Functions: 1990–1991 Research Summary of the Des Plaines River Wetlands Demonstration Project*. Chicago IL: Wetlands Research Inc. Retrieved from www.wetlandsresearch.org/projects

Smardon, R.C. (Ed.). (1983). *The Future of Wetlands: Assessing Visual-Cultural Values*. New Jersey: Allenheld-Osmun. Retrieved from www.esf.edu/via

Smolko, B., Huberd, R. & Tam-Davis, N. (2002). Creating meaningful stakeholder involvement in watershed planning in Pierce County, Washington. *Journal of American Water Resources Association*, 38, 981–994. doi: 10.1111/j.1752-1688.2002.tb05539.x

Tao, W., Bays, J.S., Meyer, D., Smardon, R.C. & Levy, Z.F. (2014). Constructed wetlands for treatment of combined sewer overflow in the U.S.: A review of design challenges and application status. *Water*, 6(1), 3362–3385. doi: 10.3390/w60x000x

Tzoulas, K., Korpela, K., Venn, S., Yli-Pelkonen, V., Kazmierczak, A., ... James, P. (2007). Promoting ecosystem health in urban areas using green infrastructure: A literature review. *Landscape and Urban Planning*, 81, 167–178. doi: 10.1016/j.landurbplan.2007.02.001

U.S. Environmental Protection Agency (USEPA). (2009). *Valuing the Protection of Ecological Systems and Services: A Report of the EPA Science Advisory Board*. Washington D.C.: U.S. Environmental Protection Agency.

U.S. Environmental Protection Agency (USEPA). (2011). Keeping Raw Sewage and Contaminated Stormwater Out of the Public's Water. New York, NY: USEPA Region 2. Retrieved from www.epa.gov/region2/water/sewer-report-3-2011.pdf

U.S. Environmental Protection Agency (USEPA). (n.d.). What is green infrastructure? Retrieved from www.epa.gov/green-infrastructure/what-green-infrastructure

Violin, C.R., Cada, P., Sudduth, E.B., Hassett, B.A., Penrose, D.L. & Bernhardt, E.S. (2011). Effects of urbanization and urban stream restoration on the physical and biological structure of stream ecosystems. *Ecological Applications*, 21(6), 1932–1949. Retrieved from www.jstor.org/stable/41416629

Walker, G.B., Senecah, S.L. & Daniels, S.E. (2006). From the forest to the river: Citizens' views on stakeholder engagement. *Human Ecology Review*, 13, 193–202. Retrieved from https://digitalcommons.usu.edu/cgi/viewcontent.cgi?referer=https://www.google.com/&httpsredir=1&article=1010&context=swa_facpubs

Wolch, J.R., Byrne, J., & Newell, J.P. (2014). Urban green space, public health and environmental justice: The challenge of making cities "just green enough." *Landscape and Urban Planning*, 125, 234–244. doi: 10.1016/j.landurbplan.2014.01.017

Yli-Pelkonen, V., Pispa, K. & Helle, I. (2006). The role of stream ecosystems in urban planning: A case study from the stream Rekolanoja in Finland. *Management of Environmental Quality: An International Journal*, 17(6), 673–688. doi: 10.1108/14777830610702511

9 Creative engagements with waterway restoration and environmental justice

Sharon Moran

Overview

Many creative projects approach degraded streams and lost waterways in provocative ways. In some cases, the justice dimensions are clearly present in what people paint on walls, stencil on streets and even perform by the water's edge. This chapter highlights some of the many works of art that simultaneously connect justice issues with waterway restoration themes.

Artists working in a range of mediums have independently produced work relating to waterways, and these often engage waterway memories as well as themes that are more conceptual, including redemption, dispossession and justice, among others. Works mentioned below include pieces that are both professional and amateur, and some are even anonymous. We acknowledge that we could include examples from virtually every art form, but for simplicity we have selected primarily examples from the visual arts.

Lots of state-funded restoration projects use the arts as an avenue for action, to help build good will among local residents. It is community members—rather than self-defined artists—who have produced some of the works discussed below. Using the arts, many restoration initiatives have invited community members (and especially youth) to respond to their once-and-future waterways. This is goal-oriented on the part of the organizers, and they explain the role of building expressive connections and how it can help build people's attachment to these waterways. Below we review a range of creative engagements, covering several pieces, which call attention to the loss, recovery and potential of degraded urban waterways and their communities.

Creative engagement examples

Murals showing California's first people, fishing and living with waterways (San Diego, CA)

San Diego's Chicano Park was established out of a struggle for self-determination, and it commemorates and celebrates Chicano

Figure 9.1 Mural by Victor Ochoa, "All the Way to the Bay," in San Diego's
        Chicano Park
Source: Photo by Jimi Giannatti

(Mexican-American) heritage. Founded by community residents in the wake of political battles for survival, this 7-acre park tells stories through a total of nearly 50 different murals constructed over the past 40 years. The park is a "barrio tragedy transformed into triumph," the founders explain, and "by claiming Chicano Park, the descendants of the Aztecs the Chicano Mexicano people begin a project of historical reclamation. We have returned to Aztlán—our home" (Anguiano, 2017, para. 2). The community, located in an area that has been bisected by a highway overpass, once encompassed many thousands of people and stretched all the way to the Bay (Warth, 2017). However, in various rounds of rezoning, residential segregation practices, restrictive covenants and expropriation, the original residents were pushed out from many areas, shrinking the size of the community and eliminating access to the Bay. Community activists are responsible for the park's existence as well as its continued maintenance; they would like to establish connections to the water once again, and this mural, emblazoned with "All the Way to the Bay" (Figure 9.1), testifies to this goal as well as to the past. Like other works there, it celebrates the society of the first people in the area, which stretched across what is now called the Southwest. Because it centers on history and waterway access, especially to the San Diego Bay, this piece involves interconnected concepts of identity, citizenship and environmental justice.

Mosaic tiles depicting anadromous fish (Syracuse, NY)

Efforts to restore Onondaga Creek (Syracuse, NY) will hopefully lead to the return of a viable population of an indigenous fish, the Atlantic salmon. This tile installation (Figure 9.2), designed by mosaic artist Jillian Hirsch, was installed in July 2017 by community volunteers. It is located on a path that winds through downtown Syracuse, highlighting a once-and-future local species, right where it once thrived. The Onondaga Environmental Institute received funding for the project from the New York State Council for the Arts. "We don't even know what we're missing, with the absence of the fish migration," said Amy Samuels, Program Manager (personal communication, December 5, 2017). One way this art connects to a justice concept relates to the Native Americans who live in the Syracuse area and to their community, the Onondaga Nation, which lies about 8 miles upstream, closer to the headwaters. This fish is an important part of the Nation's spirituality and worldview, and its return to Onondaga Creek would be a powerful and important change.

Stenciled landscape images showing regional water bodies and indigenous names (Syracuse, NY)

The images of place names shown in Figure 9.3, along with the communities and waterways they reference, are stenciled in a neighborhood in Syracuse, NY, where intersecting streets bear their names. The project, developed by Peter Edlund, calls attention to the evocative place names developed by indigenous people in the area (Xu, 2013). For example: in the Haudenosaunee language, the term *Otisco* means, "water much dried away," and the word *Wyoming* means "at the great river meadow." According to a United Nations report, "The deliberate and state-imposed destruction of indigenous languages has caused the loss of traditional knowledge systems" (Settee, 2008, p. 1), and thus the connection to identity, sovereignty and justice is clear. Few people would know that the local place name had that origin, so this project, labeled "Forgotten New York," helps call attention to the first people in the area, their knowledge system and their continued significance.

Sound walk pieces using on-location interviews (Onondaga Creek, NY)

This project, called "Up the Creek," used community conversations and individual interviews as the foundation for a sound-based walking tour. Artist Fereshteh Toosi captured people's reactions to being asked about the creek in their neighborhood, including responses that were affectionate as well as apprehensive (Rhodes, 2008). These statements and recollections were assembled into a sound piece that was placed on a device, and people can participate by listening, ideally while walking creekside (Figure 9.4). Most

Figure 9.2a and b Salmon mosaic being installed near Onondaga Creek, Syracuse, New York

Source: OEI approved photo by Amy Samuels

Figure 9.3a and b Sidewalk stencils depicting place names that were used by the
first peoples, in Haudenosaunee language

Source: Photos by S. Moran

Figure 9.4a and b Sound art project with developers and listeners, at Syracuse
University and walking along Onondaga Creek, Syracuse, NY

Source: Photos by Fereshteh Toosi

residents in the immediate area are African-American and would be counted as low income by demographers; in this work, people's comments make it clear that the degraded state of the creek is an environmental justice issue.

The Golden Ball celebration (Bronx River, NY)

With an interest in stimulating community engagement, the Bronx River Alliance commissioned two artists to develop an idea to build interest and excitement, and the first Golden Ball procession was held in 1999. Each year in April, thousands of people assemble by the water's edge in self-propelled vessels to escort this large, shiny object down the river, over 10 miles (Filip, 2014; see Figure 9.5). According to the artists, Mags Harries and Lajos Héder, the gilded orb is "a symbol of the sun, energy and spirit of the Bronx River," and its ambiguity "encourages people to invent stories about it and add their own mythology" (Harries & Héder, 1999). Along the way, a festival atmosphere prevails: people hold music and dance events, and also plant trees and run cleanup projects (Bronx River Alliance, 2017). One of the original intentions of the project developers was to help underscore continuity in an area that many see as fractured; the communities immediately adjacent to this stretch of the Bronx River vary in character, and because some are considered environmental justice communities, this project is creating intersections involving themes of waterways and justice.

Graffiti about wetland birds (Los Angeles, CA)

This mural, featuring the prominent phrase "creatures of justice," was put up during a graffiti competition held at the Arroyo Seco confluence of the Los Angeles River in 2007 (Meeting of Styles, 2007). Centering on beautiful waterfowl whose habitat has been decimated by urban development (Arroyo, 2010), this piece (Figure 9.6) makes a reference to encroachments into wetlands in a medium rarely used for this theme. It is also striking that the artist references the urbanized watershed and the role of utilities through depictions of a sewer access port and a worker in a hard hat; no detail is available about the identity of the artist or any related work. Unfortunately, this piece was relatively short-lived: just as the graffiti competition ended, the City declared that the program staff did not obtain the appropriate permits in advance, and the art's removal was ordered (Brayj, 2007; Meeting of Styles, 2007).

Participatory action and temporary installations for restoring rights of urban swimming (Paris, France)

Activists in Paris, France, are seeking to regain the right to swim in urban waterways. Their creative engagements consist of performance (such as staging public swimming events) (Figure 9.7) as well as visual representations

Figure 9.5 a) Golden Ball event on Bronx River, NY; b) Golden Ball Festival dancers
Source: a) Photo by Anne-Marie Runfola b) Photo by Mags Harries and Lajos Héder

(a)

(b)

Figure 9.6a, b, and c Graffiti in Los Angeles, CA (near Arroyo Seco) depicting
wetlands, urbanization and the slogan "Creatures of Justice"

Source: Umberto Brayj/CC BY 2.0

(c)

Figure 9.6 (Cont.)

Figure 9.7 People diving into open water in Paris, France (Bassin de la Villette), as part of a "wild swimming" event

Source: Photo by Olivier Ortelpa

of fantasy futures. One of their slogans is "Le retour de la nature en ville c'est bien, le retour de la baignade en ville c'est mieux" (the return of nature in the city is good, the return of bathing in the city is better). The group, called Laboratoire des baignades urbaines expérimentales, or Laboratory for Experimental Urban Swimming, is fairly new (started in 2013). While not uninformed about the health risks, they eschew managerial approaches and resist the authority that separates them from free swimming and unimpeded access to water (Voyer, 2017), especially in urban Paris (where heat waves have killed thousands) and particularly for minorities and immigrants (Morenne, 2017). The group celebrates swimming as a right associated with citizenship; being deprived of that right is condemned as an injustice, especially for urban residents who may not have other options. They criticize government for allowing the contamination of the waterways in the first place, and their goals include persuading government(s) to stop forbidding urban swimming and better watershed management on the part of the authorities. Their public actions and fantasy postings all seek to cultivate people's imaginations, raising the profile of the issue among the general population.

Conclusion

Across a range of mediums and forms, artists have created works that invite contemplation about people's connections to urban waterways and the concepts of (in)justice they may manifest. Taken together, these works testify to the multifaceted ways people inhabit urban waterway communities, and how they can be acknowledged, mourned, puzzled over and celebrated. These works underscore the value of creative arts in provoking reflection on possibilities, and in stimulating people's imagination in ways that transcend managerial approaches.

Although this chapter centers on visual arts, literary examples of references to urban waterways are plentiful, and they may be among the best-known creative engagements with urban waterways. Literary works are important because they are so accessible, and they help cultivate people's conceptual vocabularies about waterways. References to urban waterways have appeared in many poems and short stories in which their value as symbols and metaphors is pivotal; Langston Hughes' classic poem, "The Negro Speaks of Rivers," also engages Black American identity (Miller, 2006) and connects themes that might now be interpreted as environmental justice. Written in 1921, during the Harlem Renaissance, this work celebrates African-American identity across space and time using images of color, place and flow, and it "reclaims the origins in Africa of both physical and spiritual humanity" (Miller, 2006, p. 57).

While purposeful communication has an important place in restoration projects, these creative pieces are not necessarily conveying a message as much as inviting contemplation. They stimulate people to pause and think about these waterways as well as their connection to them, now and in the future. Encountering these works makes people wonder, "What am I looking at here?" which in turn triggers reflections on such larger questions as "What nature?" "Whose nature?" and "For what purpose?" Questions about concepts of rights and entitlements can also be raised, especially the role of the state in facilitating those rights. Responsibility and citizenship are fundamental to urban ecosystems with greater integrity, as well as communities where greater justice prevails.

References

Anguiano, M. (2017). The Battle of Chicano Park. Retrieved from http://chicano-park.com/cpscbattleof.html

Arroyo, J C. (2010). Culture in concrete: Art and the re-imagination of the Los Angeles River as civic space. (Unpublished master's thesis), Cambridge, MA: MIT. Retrieved from http://hdl.handle.net/1721.1/59727

Brayj, U. (2017). 2007 Meeting of the Styles at the LA River Arroyo Seco confluence. Retrieved December 31, 2017 from Flickr: www.flickr.com/photos/ubrayj02/6770050785/in/album-72157629056782299/

Bronx River Alliance. (2017). Launching the Golden Ball. Retrieved from http://bronxriver.org/?pg=content&p=abouttheriver&m1=13&m2=78&m3=97

Filip, A.J. (2014). Large-scale urban planning schemes in the hands of citizens: The Greenway in the New York Borough of the Bronx. *Architectural and Town Planning Quarterly, Polish Academy of Sciences, LIX* (1), 69–85. Retrieved from www.kaiu.pan.pl/images/1.2014pdf/A.J.Filip_eng.pdf

Harries, M. & Héder, L. (Producers). (1999, April 24). The Bronx River Golden Ball [Video file]. Retrieved from https://vimeo.com/51022223

Meeting of Styles. (2007, September 22–23). [Blog entry]. Retrieved from www.meetingofstyles.com/blog/22-23-september-2007-los-angeles-usa

Miller, R.B. (2006). *The Art and Imagination of Langston Hughes*. Lexington, KY: University Press of Kentucky.

Morenne, B. (2017, September 5). Public pools in Southern France become a measure of inequality. The New York Times. Retrieved from www.nytimes.com/2017/09/05/world/europe/public-pools-in-southern-france-become-a-measure-of-inequality.html?_r=0

Rhodes, N.K. (2008, March 27). Up the Creek: Public sound art project well underway. Syracuse City Eagle, p. 11.

Settee, P. (2008). *Native Languages Supporting Indigenous Knowledge*. New York, NY: United Nations Department of Economic and Social Affairs.

Voyer, A. (2017, March 9). Outdoor swimming in Paris with the canal club—in pictures. The Guardian. Retrieved from www.theguardian.com/travel/gallery/2017/mar/09/outdoor-swimming-paris-canals-river-seine-outdoor-swimming-society

Warth, G. (2017, January 11). Chicano Park named national historic landmark. San Diego Tribune. Retrieved from www.sandiegouniontribune.com/news/politics/sd-me-chicano-historic-20170111-story.html

Xu, A. (2013, August 7). Artists Focus on the Near West Side. Syracuse New Times. Retrieved from www.syracusenewtimes.com/artists-focus-near-west-side/

10 Summary and streams of revitalization practice

April Karen Baptiste, Sharon Moran and Richard Smardon

Our plan for this book is to explore the potential of learning through deliberative process (Petts, 2007) and collaborative learning models in general (Daniels & Walker, 1996) with social equity. Basic models are presented for organizing multiple stakeholders for purposes of waterway revitalization or naturalization—if not restoration. The book focuses on environmental justice issues and on case studies throughout North America and Europe. The objective is to find common lessons learned from successful and unsuccessful participatory processes for river, creek and stream revitalization within urban areas and for addressing environmental justice issues.

Chapters 1 and 2 summarize key aspects of urban waterway restoration/revitalization/naturalization history and policy as practiced in North America and Europe. The academic research community drove much of the waterway restoration science; a relatively small number of social scientists and practitioners were involved with the social side of stakeholder involvement and exploring agenda-setting practices for urban waterway revitalization. These two introductory chapters substantiate the need to incorporate social and ecological service functions with urban waterway restoration projects as well as balance such functions as part of the revitalization process. There is also a need to involve stakeholders in fact-finding and design aspects of such a process (Moran et al., 2013; Petts, 2006), thereby creating ownership and investment for stakeholders.

Chapters 3 and 4 lay out core theoretical foundations, specifically environmental justice and political ecology, which have promise for application to urban waterway restoration/revitalization issues. Although the history of environmental justice stems primarily from U.S. cases, the principles for procedural and distributive environmental justice are equally applicable to non-U.S. contexts. For procedural justice, a key question would be: How did disenfranchised urban communities become involved in waterway revitalization projects? For distributional EJ, the main question would be: Who benefits? From a political ecology standpoint, understanding who holds power and how it is manifested within socio-environmental systems is pivotal to gaining greater purchase on EJ issues regarding stream restoration. This will also help develop better perspectives on how we integrate stakeholders from under-served communities in meaningful and effective ways.

Chapter 4 makes a deeper analysis of EJ elements such as leadership characteristics, neighborhood demographics, revitalization initiation process, reasons for the process, strategies utilized and revitalization outcomes as they were manifest in five case studies: Anacostia River, Washington, D.C.; Bronx River, New York; Mill Creek, Philadelphia, Pennsylvania; Chattanooga Creek, Tennessee; and Onondaga Creek, Syracuse, New York (see Tables 4.1 and 4.2). The cases in this chapter illustrate ways to begin the conversation about how environmental justice principles are incorporated into river revitalization efforts. While EJ principles have served some communities well, there are challenges to ensuring how environmental justice is achieved. What must be noted, though, is that there are benefits to having an EJ focus on river revitalization, with the most significant being reducing the impacts of gentrification on urban communities. Given the current need for urban renewal in many spaces, and the drive for access to ecological resources in the urban context, river revitalization must continue to deliberately incorporate environmental justice principles into the planning, implementation and maintenance phases.

Chapters 5 through 9 look at waterway revitalization process models that are focused on application and implementation. There has been little theoretical or practical documentation, especially with reference to urban waterways planning processes; however, there are some emerging methods on engaging waterway communities and stakeholders in a positive process. Social learning and co-production are two keys that have been utilized in both Europe and North America. When incorporating environmental justice issues, such processes should also show interactive, distributive and deliberative justice, and new metrics may be needed to evaluate prospective projects (Moran, 2007).

Chapter 5 includes a review of waterway organizational schemas for some six different projects in the U.S. This review includes public–private partnerships, not-for-profits that bridge over time with different city or county administrations; urban land trusts and similar organizations that raise money, acquire land and/or easements and pass them on to public agencies; and groups that organize events and activities leading to more consciousness-raising and use of urban waterways. The review also included organizations that perform many functions and some organizational schemas that separate functions among many groups. A potential research challenge is to more formally diagram and assess these organizational relationships and functions (Felleman, 1997; Väntänen & Marttunen, 2005).

A statement was made in Chapter 1 that many communities have been disconnected from urban waterways, have community environmental justice issues and in some cases hold negative views of urban waterways. Chapter 6 covers ways to reconnect these place-based narratives to address environmental and social grievances and incorporate them into urban waterway revitalization. Chapter 6 also reviews the appropriate literature, including the Mill River, Philadelphia and South Bronx case studies, and the roles elders can play in restoring urban waterways. To help unpack the complex identities of older adults working in this realm, the second part of Chapter 6 explores the role(s) of elders in environmentalism, including their contributions to activism generally. For some elders, it appears that working on restoration initiatives is more than simple voluntarism; it is rather an act that has specific significance relative to values, ethics, justice and memory. The chapter concludes with some recommended measures to reconnect valued relationships with urban waterways.

Chapters 5, 6 and 7 create the foundation for more utilization of interactive participatory methods to engage urban waterway neighborhoods in collaborative and social learning processes as part of the revitalization waterway planning (Petts, 2006, 2007; Smardon et al., 1996). Community engagement and mapping is needed to incorporate risk from water quality threats and flooding, perceived value(s), perceptions of existing waterway qualities and understanding of hydrologic and ecological processes. Chapter 7 presents more collaborative approaches, such as participatory GIS, agent-based modeling, participatory design and crowdsourcing. A participatory interactive creek revitalization design charrette process is presented as a case study that can compile specific projects by creek segment as well as project prioritization for implementation.

Revitalizing urban waterways can be considered as providing green infrastructure (GI) for urban areas, through strategies including flood amelioration, urban runoff reduction, water quality improvement and urban open space access (Barnhill & Smardon, 2012; Baptiste et al., 2015). One measure of the benefits of urban waterway GI is assessing ecosystem service improvement for urban residents. Chapter 8 includes a review of ecosystem services that urban waterways can provide with an emphasis on cultural services, ways of assessing ecosystems service and some case studies of applied ecosystem service assessment of benefits as well as environmental justice impacts to urban waterway communities. The Mayes Brook, East London, case study (Everard, 2012) is utilized as an example for quantifying such ecological services of some £880,000 annually. The Rekolanoja Finland Case Study (Yli-Pelkonen et al., 2006) was used to illustrate quality of life (cultural ecosystem services) from urban stream improvement projects. The Onondaga Creek, Syracuse, NY, case study was utilized to highlight the ways GI can be used to help address combined sewer overflows and by extension some historical environmental injustices.

Chapter 9 highlights a range of artistic pieces that can illuminate people's connections to urban waterways and the concepts of (in)justice that they may manifest. These creative works underscore the value of waterway restoration by stimulating people's imagination and provoking reflection on possibilities in ways that transcend technical managerial approaches. We provide a number of examples of use of different mediums. Concepts of rights and entitlement are also raised, especially the roles of government and other organizations in facilitating those rights.

Besides this recap of all previous content within this book, there may be some future scholarship and practical measures that would improve consideration of addressing and ameliorating EJ issues as part of urban waterway restoration.

Here are some things to consider for both future research and implementation with sensitivity to environmental justice and political ecology:

1. *How are waterways chosen for revitalization?* This question highlights distributive injustice elements in that some areas are prioritized over others. There must be a deliberate attempt to prioritize spaces where marginalized communities live to ensure that injustices are not replicated.
2. *Who is involved in the revitalization process?* Once spaces are chosen to be revitalized, procedural justice indicates that all vested stakeholders should be part of the process. Decision-makers will avoid conflict if practices of procedural justice are incorporated.
3. *More empirical research examining the ways in which environmental justice principles have been incorporated into urban river revitalization is needed.*
4. *In the same vein, we also need river revitalization efforts that build on EJ principles to be undertaken by local and state governments.* This can also be aligned with research, which can assess the effectiveness of these projects over time and the value of incorporating EJ principles into revitalization efforts.

The U.S. EPA's Urban Waters Program was established in 2012 with the goal of "seeking to help communities—especially underserved communities—as they work to access, improve and benefit from their urban waters and the surrounding land" (USEPA, 2017). The program has already supported work in more than two dozen urban waterways, and made some headway in the task of integrating environmental justice concerns into restoration efforts. This example of a program supporting urban waterway restoration while simultaneously emphasizing underserved communities demonstrates that there is now some precedent for a government program combining these goals. Together with other positive stories shared in this book, that program helps support our optimism about environmental justice goals for urban waterway restoration in North America, Europe and elsewhere, in connection with sustainability, human health and quality of life improvement.

References

Baptiste, A.K., Foley, C. & Smardon, R.C. (2015). Understanding urban neighborhood differences in willingness to implement green infrastructure: A case study of Syracuse, NY. *Landscape and Urban Planning, 136*, 1–12. doi: 10.1016/j.landurbplan.2014.11.012

Barnhill, K. & Smardon, R.C. (2012). Gaining ground: Green infrastructure attitudes and perceptions from stakeholders in Syracuse, NY. *Environmental Practice, 14*(1), 6–16. doi: 10; 1017/s1466046611000470

Daniels, S.E. & Walker, G.B. (1996). Collaborative learning: Improving public deliberation in ecosystem-based management. *Environmental Impact Assessment Review, 16*(2), 71–102. doi: 10.1016/0195-9255(96)00003-0

Everard, M. (2012). What have rivers ever done for us? Ecosystem services and river systems. In B.J. Boon & P.J. Raven (Eds.). *River Conservation and Management* (pp. 313–321). London: Wiley-Blackwell.

Felleman, J. (1997). *Deep Information: the Role of Information Policy in Environmental Sustainability*. Greenwich, CT: Ablex.

Moran, S. (2007). Stream restoration projects: A critical analysis of urban greening. *Local Environment, 12*(2), 111–128. doi: 10.1080/13549830601133151

Moran, S., Perreault, M. & Smardon, R.C. (2013). Finding our way: Urban waterway restoration and participation process. In J.G. Fabos, M. Lindhult, R.L. Ryan & M. Jackson (Eds.). *Proceedings of Fabos Conference on Landscape and Greenway Planning: Pathways to Sustainability* (pp. 20–35). Amherst, MA: University of Massachusetts.

Petts, J. (2006). Managing public engagement to optimize learning: Reflections from urban river restoration. *Human Ecology Review, 13*(2), 172–181.

Petts, J. (2007). Learning about learning: Lessons from public engagement and deliberation in urban river restoration. *The Geographical Journal, 173*(4), 300–311. doi: 10.1111/j.1475-4959.2007.00254.x

Smardon, R.C., Felleman, J.P. & Senecah, S.L. (1996). *Protecting Floodplain Resources: A Guidebook for Communities*. Washington, D.C.: Prepared for the Federal Interagency Floodplain Management Task Force, U.S. Government Printing Office.

U.S. Environmental Protection Agency (USEPA). (2017). About the Urban Waters Movement. Retrieved from www.epa.gov/urbanwaters/about-urban-waters-movement.

Väntänen, A. & Marttunen, M. (2005). Public involvement in multi-objective water level regulation development projects—evaluating the applicability of public involvement methods. *Environmental Impact Assessment Review, 25*, 281–304. doi: 10.1016/j.eiar.2004.09.004

Yli-Pelkonen, V., Pispa, K. & Helle, I. (2006). The role of stream ecosystems in urban planning: A case study from the stream Rekolanoja in Finland. *Management of Environmental Quality: An International Journal, 17*(6), 673–688. doi: 10.1108/14777830610702511

Index